Elements of
Machine Design

EMANUEL ROSENTHAL

and

GEORGE P. BISCHOF

McGRAW-HILL BOOK COMPANY, INC.   1955

New York   Toronto   London

ELEMENTS OF MACHINE DESIGN

Library of Congress Catalogue Card Number: 54-11765

9 10 11 – M P – 9 8 7

53835

PREFACE

The purpose of this book is to analyze the problems inherent in the proportioning of machine parts and to present simple and rational solutions. Designed primarily for use in technical institutes and technical high schools, the text will also prove helpful to students on other levels of instruction for a fuller understanding of basic principles and certain short cuts in design practice.

Although a knowledge of strength of materials is assumed, many of the fundamental concepts of this subject are reviewed before applications are made to machine design. Emphasis is placed on the derivations of the formulas presented and the relationship of these formulas to the principles of strength of materials on which they are based. Many of the formulas are special in nature and consist of terms applicable only to the particular phase of machine design under discussion. This is as it should be, for the subject itself is a specialized branch of mechanical engineering, and when the student of machine design has mastered basic principles he turns, as does the professional machine designer, to a special and specific formula for help in his special problem.

The problems at the end of each chapter are practical, comprehensive in scope, and graded in difficulty.

The instructor may wish to omit certain topics and certain problems as either too elementary or too advanced. Likewise, various schools are bound by differing syllabi and have different prerequisites for the subject. For example, where the subject of applied mechanics is a prerequisite, Chaps. 4 and 5 may seem superfluous. In cases where omissions are found desirable, they can be made without any serious break in the continuity of the text.

The authors are indebted to Arthur M. Shrager of Brooklyn Technical High School for help on properties and heat-treatment of metals and to Howard G. Schneider of Brooklyn Technical High School for help in certain phases of applied mechanics.

Emanuel Rosenthal
George P. Bischof

CONTENTS

Preface

1. The Work of the Machine Designer   1
2. Properties of Materials   7
3. Metals of Industry   17
4. General Problems of Force and Motion   27
5. Force and Motion as Applied to Simple Machines   40
6. Design Stress and Dynamic Loading   50
7. Power and Power Transmission   58
8. Shafts in Torsion Only   66
9. Shafts in Bending and Torsion   73
10. Elastic Deformation in Shafts   86
11. Keys   93
12. Pulleys and Belts   101
13. Gears and Friction Wheels   112
14. Couplings   128
15. Bearings   135
16. Clutches   145
17. Thin-walled Cylinders; Welded and Riveted Joints   154
18. Screws, Fastenings, and Seals   169
19. Springs   181
20. Combined Stresses   191
21. Fits, Allowances, and Tolerances   208

Appendix: Areas and Moments of Inertia about Center-of-gravity Axes of Simple Geometric Sections   215

Table of Trigonometric Ratios   216

Bibliography   223
Visual Aids   225
Index   229

CHAPTER 1

THE WORK OF THE MACHINE DESIGNER

1. What Is Machine Design? Examine the different mechanisms illustrated in Fig. 1. You will see that in some cases they change the *rate* or *direction* of motion (for example, in *b* and *d*), whereas in other cases the *type* of motion is changed as from rotary to straight line (for example, *i* and *j*). The motion of the wheel *B* in *k* is a combination of both types, for a downward pull on chain *A* causes wheel *B* to move upward as well as to rotate counterclockwise. A mechanism can be defined as a device consisting of two or more parts, one of which in moving causes motion in the others in a definite relationship to the motion of the first part.

Many mechanisms are applications of the simple machines studied in physics, namely, the lever, wheel and axle, inclined plane, and pulley. Some are used primarily to increase the force; others are used to increase the speed. When one of the two (speed or force) increases, the other necessarily decreases. Try to identify the simple machine which is the basis of each of the mechanisms illustrated. The more complex machines used in factories are nothing more than combinations of mechanisms which in turn are applications of the simple machines.

If any part of a machine breaks, the whole machine must be taken out of service while replacement is made. Therefore, when a new machine is being designed, it is of the utmost importance that each part be made sufficiently strong to prevent failure. On the other hand, if the parts are designed to withstand much greater loads than will at any time be applied, the manufacturing costs become excessively high and the whole machine needlessly large and bulky. Also, most machines are limited in size by the use to which they will be put. A revolver should be the proper size for the hand, and a sewing machine the right size for the operator and the cloth to be sewn. As in all cases of engineering design, the design of machines is largely a balance between strength and economy, within the proper size limitations.

1

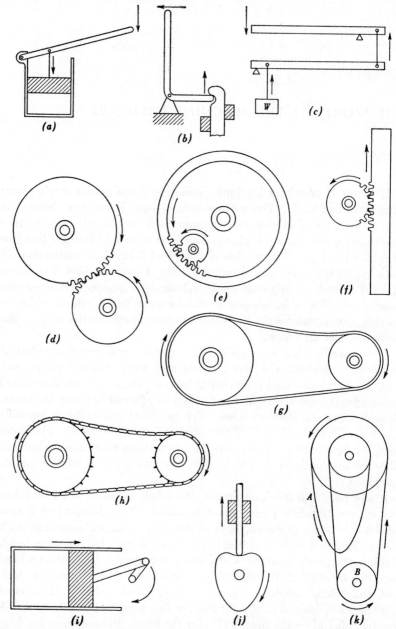

Fig. 1. Some common types of mechanisms.   (a) Simple lever; (b) bent lever; (c) compound lever; (d) mating gears—external contact; (e) mating gears—internal contact; (f) rack and pinion; (g) pulleys and belt; (h) sprockets and chain; (i) gas engine; (j) cam and follower; (k) differential hoist.

The work of the machine designer involves the selection of the most suitable mechanisms and the selection of the proper materials and proportions for the parts.

2. Steps in Machine Design.   After the machine designer has come to a decision as to the type of mechanism to be used, the next step is to determine as accurately as possible the nature, magnitude, direction, and point of application of each of the forces acting on the parts.   For this the designer must have a working knowledge of certain laws of physics.   Then with his knowledge of the properties of materials, he is able to make intelligent choices of the most suitable materials to use, and finally, with his knowledge of strength of materials, he is able to make calculations to determine the proper proportions of the parts. The following chapters deal with those phases of machine design *after* the selection of the particular mechanisms has been made.   In other words, we will concern ourselves with the analysis of the forces, the selection of the materials, and the design of the details of those parts commonly used in modern machinery.

3. Engineering Design—an Inexact Science. In attempting to make calculations for proportioning an engineering structure, the designer is plagued by two factors of fundamental importance.

1. It is usually impossible to determine fully the characteristics of the loads applied to a body.   This is true in the design of stationary structures as buildings and bridges, but is especially true of machines where loads cause motion of the parts.

2. Samples of a given material vary in their ability to resist loads. Two finished products made from the same type of steel, for example, may show variation in strength owing to some hidden flaw in one of them.

Because of these two considerations engineering design can never rise to the status of an exact science.   This does not mean, however, that design calculations should be thrown overboard and that proportioning be done by mere guesswork or rule of thumb.   Using certain assumptions for the unknowns as a basis, calculations do result in a proportioning of parts between the extremes of those excessively costly and those which may break under load.   There is, therefore, a need for more, not less, calculation in design.

4. Use of Manufacturers' Catalogues.   When the mechanism consists of one or more standard machine parts, such as pulleys or gears, the designer will frequently select a ready-made part from a manufacturer's catalogue.   As an illustration of the information contained in these catalogues, a portion of a page of one is reproduced in Fig. 2. The sizes of the pulleys range from $1\frac{1}{2}$ in. to 14 in. outside diameter

Chicago SINGLE GROOVED PULLEYS for ("A"-Section (½" x 11/32") Belts)

KEYWAY DATA

Bores indicated by (•k) have standard keyways.

Bores 5/8"-3/4"-7/8" have 3/16"x3/32" K.W.
Bores over 7/8" have 1/4"x1/8" K.W.

All pulleys are furnished with slotted headless set screws. Pulleys with special bores, keyways, and set screws can be had at a slight increase in price.

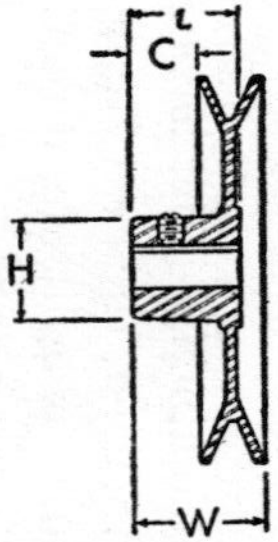

Stock No.	Out-side Diam.	Pitch Diam.	Price	Diameter of Bore 3/8"	1/2"	5/8"	3/4"	1"	1 1/4"	W	C	L	H	No. of Spokes	Net Weight
120	1½"	1.25"		•	•					1⅛"	½"	1"	⅞"	None	2.5 ozs.
127	1¾"	1.5"		•	•	•	•			1⅛"	½"	1"	1 1/16"	None	4.0 ozs.
121	2"	1.75"		•	•	•	•k			1⅛"	½"	1"	1⅛"	None	4.75 ozs.
123	2¼"	2"		•	•	•	•k			1⅛"	½"	1"	1⅛"	None	5.75 ozs.
106	2½"	2.25"		•	•	•	•k			1⅛"	½"	1"	1⅛"	None	6.5 ozs.
124	2¾"	2.5"		•	•	•	•k			1⅛"	½"	1"	1⅛"	None	7.5 ozs.
108	3"	2.75"			•	•k	•k	•k		1⅛"	½"	1⅛"	1¼"	None	9.75 ozs.
126	3¼"	3"			•	•k	•k			1 3/16"	½"	1⅛"	1¼"	None	10.0 ozs.
122	3½"	3.25"			•	•k	•k			1¼"	⅝"	1⅛"	1¼"	None	11.25 ozs.
109	4"	3.75"			•	•k	•k	•k		1¼"	⅝"	1⅛"	1¼"	None	13.25 ozs.
128	4½"	4.25"			•	•k	•k			1¼"	⅝"	1⅛"	1¼"	None	16.0 ozs.
135	5"	4.75"			•	•k	•k	•k		1⅜"	⅞"	1 3/16"	1 9/16"	Four	14.25 ozs.
134	5½"	5.25"			•	•k	•k	•k		1½"	⅞"	1 5/16"	1 9/16"	Four	16.0 ozs.
136	6"	5.75"			•	•k	•k	•k		1½"	⅞"	1 5/16"	1 9/16"	Four	18.0 ozs.
137	7"	6.75"			•	•k	•k	•k		1½"	⅞"	1 5/16"	1 9/16"	Five	22.0 ozs.
138	8"	7.75"			•	•k	•k	•k		1½"	⅞"	1 11/32"	1⅜"	Five	25.0 ozs.
1009	9"	8.75"			•	•k	•k	•k	•k	1½"	13/16"	1 11/32"	2 1/16"	Six	35.0 ozs.
1010	10"	9.75"			•	•k	•k	•k	•k	1½"	13/16"	1 11/32"	2 1/16"	Six	38.0 ozs.
1012	12"	11.75"			•	•k	•k	•k	•k	1½"	¾"	1 11/32"	2 1/16"	Six	47.0 ozs.
1014	14"	13.75"			•	•k	•k	•k	•k	1½"	¾"	1 11/32"	2 1/16"	Six	54.0 ozs.

FIG. 2. Information supplied in manufacturers' catalogues. (*Chicago Die Casting Mfg. Co.*)

(OD) with varying diameters of the hole for the shaft.   For each size the principal detail dimensions are given to aid the designer.

The selection of a standard-sized part from a manufacturer's catalogue makes for time saving and for economy in the cost of the part.   In this practice the proper proportioning of the part for strength and economy according to the principles of strength of materials is not eliminated, but is shifted to the manufacturer's engineering department.   A manufacturer cannot afford to offer to prospective customers merchandise which does not meet exacting tests of both strength and economy.   Nevertheless, it is a wise practice for the customer to check the strength by calculation before purchase.   Some manufacturers' catalogues contain considerable design information.

If, on the other hand, many identical parts are needed, that is, the machine is to be reproduced in quantity and on a production basis, the

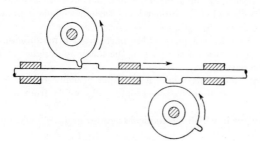

Fig. 3. An original mechanism.

designer might very likely discover by calculation that a part less heavy and bulky than a standard-sized one is sufficiently strong.   In such cases he might be able to save money by ordering the part especially made to his design.

For the design of original mechanisms where odd-shaped parts are to be used, manufacturers' catalogues are of no value.   Such parts should be proportioned by calculation and must be especially made. A mechanism of this type is illustrated in Fig. 3, where the wheels impart an intermittent back-and-forth motion to the bar.

5. Use of Engineering Handbooks.   Engineers engaged in the work of machine design find mechanical-engineering handbooks most valuable ready sources of reference.   Handbooks differ from textbooks in that handbooks cover a wide range of related fields with the information concisely presented, whereas textbooks are limited in scope and emphasize explanation of principles.

The sections in mechanical-engineering handbooks dealing with machine design present many specific formulas for use in practical problems of proportioning machine parts. Numerous charts and graphs based on the formulas present a picture of the relationship of the terms of the formulas involved, as, for example, a graph showing the required size of a shaft for varying speed and horsepower transmitted. From such a graph a designer is able to read directly the value of the unknown and sought-for quantity, thus eliminating mathematical computation. These handbooks also furnish numerous tables, such as those giving sizes of standard screw threads, sizes of keys, and the like. Finally, these handbooks, the various sections of which are written by experts in their respective fields, contain a wealth of practical design information and engineering "know-how" not ordinarily found in textbooks.

PROBLEMS

1. Why is it necessary for a machine designer to have a good knowledge of strength of materials and of the properties of various engineering materials?

2. Why is machine design said to be an inexact science?

3. How does the selection of machine parts from a manufacturer's catalogue make for a saving of time on the part of the machine designer and for economy in the cost of the part?

4. Under what conditions might a machine designer decide to have a part especially made rather than to select one ready made from a manufacturer's catalogue?

5. Name the ways in which mechanical-engineering handbooks are useful to the machine designer.

6. Outline the necessary steps to be taken by a designer who is given the job of designing a new type of machine. Carry the outline up to the point of making shop drawings.

CHAPTER 2

PROPERTIES OF MATERIALS

6. Materials and Their Properties. When the machine designer knows the characteristics that are desired of a material for a machine part, and knows the degree to which various materials possess these characteristics, he is able to make an intelligent choice of the most suitable material to use. The behavior characteristics of a material are known as its *properties*. In this chapter we shall discuss briefly the most important properties of materials from the point of view of machine design.

7. Stress and Strain. Strictly speaking, the term *stress* means the total resistance that a material offers to an applied load. However, as commonly used in engineering work, the term stress means *unit stress*, that is, the resistance that the material offers over a unit of its stressed area, and stress is therefore expressed in terms of pounds per square inch (psi). In this book this generally accepted practice is followed and it is to be understood that by stress, unit stress is meant. Also to designate the total resistance of a body, the term total stress is used.

Likewise, strictly speaking, *strain* means the total deformation measured in the direction of the line of stress. Again this term is generally used in the sense of *unit strain*, that is, the amount of deformation per unit of length, and is expressed in terms of inches per inch (in. per in.). In this book this generally accepted practice is also adhered to and it is to be understood that the term strain means unit strain.

8. Tension Test of Steel. Certain properties of materials are revealed by performing laboratory tests, such as tension, compression, and bending. As a background for the discussion of properties, let us review the tension test on a standard steel specimen. The ends of the specimen (Fig. 4) are threaded to fit the special screw grips of the machine. The shank is circular in cross section and machined to 0.505 in. diam. This odd diameter is chosen because the cross-sectional area then figures to be almost exactly 0.2 sq in. An instrument called an extensometer is clamped on the shank, and by means of this device the amount of stretch can be read, usually to ten-thousandths of an inch. If the upper and lower clamps of the extensometer are 2 in. apart, each reading is the elongation in 2 in. of length.

The first two columns of Table 1 show loads and extensometer readings for every thousand up to 5000 lb. The third column is the stress s, expressed in units of pounds per square inch (psi), and calculated for each load P by

$$s = \frac{P}{A} \tag{1}$$

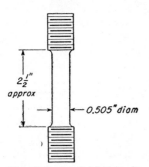

FIG. 4. A standard thread-end specimen.

where A is the cross-sectional area. Column 4 is the deformation (elongation in this case) per unit of length, that is, the strain δ, and for each extensometer reading Δ,

$$\delta = \frac{\Delta}{\text{orig. length}} \tag{2}$$

The original length in our example is the original distance of 2 in. between extensometer clamps.

Figure 5 represents a graph plotted from the values in columns 3 and 4, with stress laid off on the vertical and strain on the horizontal axis. Note that the path of the line is somewhat jagged, but when considered as a whole is straight. Inaccuracies in reading and lack of sensitivity of the extensometer account for the jaggedness.

TABLE 1

Load, lb	Extensometer readings, in. per 2 in.	Stress, psi	Strain, in. per in.
1000	0.0003	5,000	0.00015
2000	0.0006	10,000	0.00030
3000	0.0010	15,000	0.00050
4000	0.0013	20,000	0.00065
5000	0.0016	25,000	0.00080

The ideal stress-strain diagram for steel plotted from the beginning to the breaking of the specimen resembles Fig. 6. Here the first part of the graph is a straight line rising at a sharp angle. If the load were removed at any time during this first portion of the test, the extensometer pointer would return to the zero mark. This means that the specimen also has returned to its original length and as far as we can tell is perfectly elastic. Note also that the values of stress and strain are proportional for this part of the curve, as for example,

$$\frac{0.0004}{0.0008} = \frac{12,000}{24,000}.$$

After, however, the line of the graph starts to bend, the proportionality ceases, and, if the load is then removed, the specimen will no longer return to its original length, although it will return somewhat. For these reasons point A is appropriately called the *proportional*, or *elastic*, *limit*. The elastic limit of a material may then be defined as the highest stress that the material can develop and still return to its original size and shape when the load causing such stress is removed.

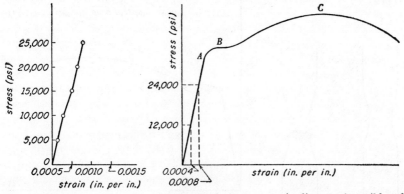

FIG. 5. Partial stress-strain diagram.

FIG. 6. Typical stress-strain diagram for mild and medium steels.

Immediately after the elastic limit, the line flattens out for a short distance (B in Fig. 6). There is little or no change in stress but considerable increase in strain. The machine is continuing to stretch the specimen but the pointer on the load dial registers very little if any change. The name *yield point* is given to the stress during this part of the test. As the test progresses, the diagram again shows an increase of stress up to the highest point. The line for this portion is curved, indicating that stress and strain are not proportional. The rate of stress increase becomes less and less as shown by the flattening of the line and the slowing down to a stop of the pointer on the load dial of the machine. *Ultimate strength* is the name given to the maximum stress (point C). From this point on, the line of the diagram drops, which means that the specimen can no longer offer as much resistance as before. Also the pointer of the load dial begins to move back from its most forward position and the specimen itself may begin to show a marked change in appearance. Somewhere along its length a thinning of the cross section, called *necking down*, may be noticed, at which point rupture finally occurs (see Fig. 7).   Necking down is peculiar to the soft and medium steels.

The stress calculations throughout are based on the original area. This gives untrue values of stress, particularly in the latter stages of the test.   Figure 8 represents two stress-strain diagrams, each plotted from the same axes and for the same specimen.   The line plotted from calculations using the actual area at no point drops back, which shows that, in reality, the specimen always takes on more and more stress,

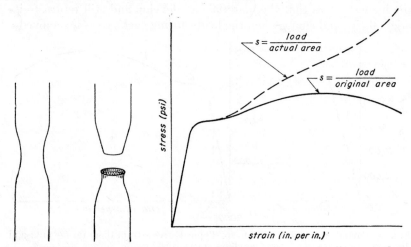

Fig. 7. Necking down and   Fig. 8. Variation in stress-strain diagram when
cup-and-cone fracture.   actual area is used.

even when necking down, until rupture.   However, in practice, the original area is always used.   Stress-strain diagrams for tests of compression, shear, and bending are similar in appearance to those for tension.

As we will recall from strength of materials, the ratio of stress to strain for any value up to the elastic limit is known as the modulus of elasticity E.   In symbols,

$$E = \frac{s}{\delta} \tag{3}$$

For values indicated in Fig. 6,

$$E = \frac{12,000}{0.0004} = 30,000,000*$$

* The units for modulus of elasticity, strictly speaking, are pounds per square inch.   Since, however, it is not a real stress, the units are omitted in this text as meaningless.

and

$$E = \frac{24,000}{0.0008} = 30,000,000$$

Because of the proportionality existing between stress and strain, any set of values below the elastic limit can be taken to calculate the value of the modulus of elasticity. Figure 9 represents the beginning of stress-strain diagrams for steels of varying strengths. Note that the line of each diagram is at the same slope, because the modulus of elasticity values vary very little. Note that the elastic limit values differ considerably.

9. Properties for Stresses below the Elastic Limit. Materials show certain characteristics below the elastic limit (elastic stress) and other characteristics above this stress. These two classes of properties will be discussed in the order named.

Elasticity is the ability to return to original shape and size. Although all materials are somewhat elastic, as far as we know, steel is perfectly elastic within the elastic stress.

Stiffness is the ability to resist deformation. All materials deform somewhat when stressed, some more, some less. For any given stress (up to the elastic stress), the material that deforms least is the stiffest.

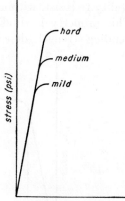

Fig. 9. Partial stress-strain diagrams for different types of steel.

From the expression $E = s/\delta$, it can be seen that E varies inversely with δ, that is, if two materials are equally stressed, the one showing the less strain has the greater modulus of elasticity. But the one with the less strain is also the stiffer. Therefore, it can be said that modulus of elasticity is a measure of stiffness, not elasticity as the name implies. Stiffness is of great importance in materials used for certain machine parts, for example, for shafting, as will be discussed later under this topic.

Resilience is the ability to deform greatly under heavy load and to spring back to the original size and shape. The modulus of *resilience* is the work done on a 1-in. cube specimen (or the energy stored by this specimen) in deforming it to its elastic limit. Work, it will be recalled from physics, is the product of the force and the distance through which this force acts. In stretching a specimen to its elastic limit in a tension test, the average force required is one-half of the elastic force (since force is applied gradually, starting at zero and ending with the elastic force as a maximum). For a cross-sectional area of 1 sq in., the force equals the stress, and for 1 in. of length, the elongation equals the strain. Hence, the work done on such a cube or its modulus of resilience is

$$\frac{\text{Elastic stress}}{2} \times \text{elastic strain}$$

As can be seen from Fig. 10, this value is equal to the area of the hatched triangle.

10. Properties for Stresses above the Elastic Limit. *Strength* is the ability to resist, without rupture, loads causing various types of stress. This property is subdivided into tensile, compressive, shearing, and bending strength, depending on the type of stress induced by the loads.

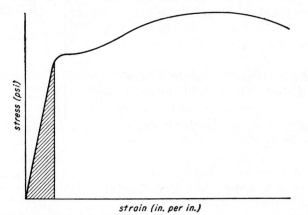

FIG. 10. Modulus of resilience from the stress-strain diagram.

Toughness is the ability to deform rather than fracture under heavy loads. This property is often important for reasons of safety when a part is subjected to an overload. The American Railway Engineering Association specifies that track bolts must be of sufficient toughness to be cold-bent 180° without showing signs of fracture.

Brittleness is the ability to fracture rather than deform. This property in a material is undesirable. As defined, brittleness is the opposite of toughness, but is more likely to be associated with suddenly applied loads and with impact loads (loads caused by one body striking another).

Ductility is the ability to be stretched (elongated). Both the percentage elongation and the percentage reduction in cross section of a specimen calculated from a tensile test are indications of the ductility of the specimen. Ductility is a special case of toughness.

Malleability comes from a word meaning "hammer" and, in a narrow sense, means the ability to be hammered out into thin sheets. As used today, it refers to the ability of a material to be worked as through forging or rolling. The properties of malleability and toughness are closely related in meaning.

Hardness is the ability to resist abrasion (wearing down) or indentation. This property is very important in the selection of materials for parts which rub on one another. The term *softness* is used as the absence of hardness.

Machinability is the ability to allow portions of the material to be removed by a cutting tool. For example, if a piece of metal can readily be turned on a lathe, it possesses a high degree of machinability.

11. Fatigue Strength. Fatigue strength is of great importance in machine design where moving parts are involved. Sometimes a material will rupture after it had at no time been stressed to its ultimate strength or even to its elastic limit. An explanation of this strange behavior lies in the manner in which the loads were applied. It is safe to say that there was either a very great number of load applications and removals or a very great number of load reversals. Figure 11 illustrates a condition of load reversals on *one* bar of metal, causing bending to the left in *a* and to the right in *b*. As we know from strength of materials, there is developed a stress of tension on the outside of the arc and one of

(a)   (b)

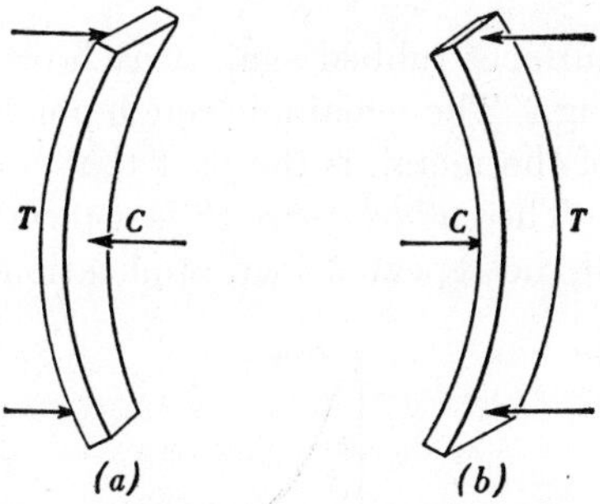

Fig. 11. Load reversals induce stress reversals.

compression on the inside (*T* and *C* in the figure). Hence, as the bending changes direction, the type of stress reverses itself. This is indicated by the exchanged positions of *T* and *C* in *a* and *b*. Although during the process of many repeated bendings, first one way and then the opposite, the stress at no time is as great as the elastic stress, yet failure might occur.

The rupturing of the material under the loading conditions just described starts by a cracking at the surface near the mid-section where the stress is greatest. As the load applications or reversals are continued, this fractured portion gradually increases in size. When the cross section has been so reduced that it can no longer resist the loads, a sudden break occurs. The initial fracture takes place where there is some surface defect as a small cut or scratch. *Fatigue strength* can, therefore, be defined as the ability to resist progressive fracturing, starting at the surface and working its way inward, under conditions of repeated load applications or reversals.

Figure 12 illustrates the cross section of a coil spring after fatigue failure from a very great number of reversals of torsional stress. Near the circumference the metal is quite smooth because there the fractured

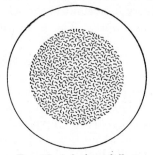

FIG. 12. A fatigue failure.

surfaces rubbed against each other while the spring was still functioning. The remaining rough portion, indicating the crystalline structure of the metal, is the part that broke suddenly.

The *endurance limit* is a term used to mark the fatigue stress limit. If the repeated load applications or reversals never induce a stress in

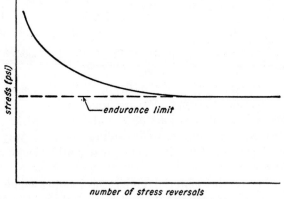

FIG. 13. A fatigue-failure diagram.

the material greater than the endurance limit, such repetitions can be made an infinite number of times without causing failure. Whereas if the stress induced is more than the endurance limit, failure will eventually occur. In the graph of Fig. 13, the relationship of stress to number of reversals necessary to cause rupture is plotted. Note that above the endurance limit the number of reversals bear an inverse relationship to the stress. This graph is typical for steel. Some nonferrous (noniron) metals have no endurance limit, meaning that no matter how small the stress induced, a sufficient number of repetitions will always cause rupture.

12. Effect of Temperature on Properties; Creep. In stating that a material possesses a certain degree of ductility or toughness, we imply that such is the case at ordinary or room temperatures. At depressed or elevated temperatures, properties of materials change in varying degrees. For example, steel becomes more brittle (less tough) at low temperatures, as demonstrated by impact tests. In a test of this kind a short-length specimen is subjected to a blow by a falling pendulum. The blow may either be applied transversely as in a beam (Charpy or Izod test) or longitudinally (tension test). At about $-40°F$ plain carbon steels become so brittle that their impact strengths are only about 15 per cent of the values at $70°F$. The addition of certain alloys to carbon steel, as, for example, nickel, to a large measure preserves the toughness at the low temperatures.

At elevated temperatures also most metals show a deterioration of their desirable properties. For temperatures higher than about $500°F$, a marked decrease in ultimate strength takes place. Thus the ultimate strength of medium cold-rolled steel at $900°F$ is about one-half of its value at $500°F$. Nonferrous metals show a less rapid decline.

In addition, because of the accompanying decrease in the elastic limit, a certain stress in a material, well below the elastic stress at room temperature, may be above the elastic stress at an elevated temperature. When this occurs, the material will permanently deform and continue to do so at a very slow rate as long as the load is applied. This plastic flow in a material is known as *creep*. The *creep strength* of a material is its ability to resist plastic flow.

In industry where low temperatures prevail, as in refrigeration plants, or where high temperatures are developed, as in the steam boilers and turbines of power plants, the effects of temperature on properties cannot be neglected in design. Furthermore, the high temperatures developed by jet and rocket engines and the temperature extremes which are the result of high-speed and high-altitude flying make this subject of very great and increasing importance. Much is still to be learned.

PROBLEMS

1. The maximum load that a material can carry rather than the load at rupture is usually referred to as the breaking load. Explain.

2. Why can any value of stress below the elastic limit and the corresponding value of strain be used to calculate modulus of elasticity?

3. If the modulus of elasticity values of steel and cast iron are 30,000,000 and 12,000,000, respectively, to what extent is steel the stiffer material?

4. In terms of magnitude of load, amount of deformation, and ability to return to original shape and size, explain how stiffness, resilience, and elasticity differ.

5. A steel specimen has a cross section in the shape of ½-in. square. The elastic load was reached at 15,000 lb and at this load a 2-in. extensometer read 0.0042 in. Calculate the modulus of resilience of the specimen.

6. Name what you consider the three most important properties that a material which is to be used as a machine part should possess.

7. One material has considerably more toughness than another. How would you expect it to act (a) in a tension test and (b) in a bending test?

8. Why is brittleness an undesirable property, especially for materials to be used as machine parts?

9. Why is fatigue strength of far greater importance in machine design where moving parts are involved than in structural design where the parts are stationary?

10. What is creep? What is its cause?

CHAPTER 3

METALS OF INDUSTRY

13. Alloys. Pure metals, such as iron, copper, or aluminum, are too soft and weak to be successfully used in modern industry as parts of structures or machines. However, when two or more metallic elements are mixed together in the molten state, there results upon solidification a metal called an *alloy* which is often much harder and stronger and, hence, far more suitable for industrial uses. Cast iron, steel, brass, bronze, and many other metals in common use are alloys. Commercial copper, often thought of as a pure metallic element, also contains small amounts of alloying elements as hardeners. The proportions of the alloying elements have considerable effect on the properties of the metal, but more important in many cases is the effect of cold working (as cold rolling or cold forging) and of heat treatment.

14. Cold Working and Heat Treatment. *Hardening.* The term hardening implies an increase in the ultimate tensile strength and brittleness, as well as in the hardness. Most alloys show such changes after being cold-worked, giving rise to the term, *work hardening.*

In addition, cast iron, steel, and some nonferrous alloys can be hardened by heating and sudden cooling as by quenching in water or oil. Some metals when heated change the form of their crystalline structure upon reaching a certain temperature. Again on cooling the change is made in reverse, but this time at a slightly higher temperature. These temperatures are called the *critical points* or *critical temperatures* and the range between them is referred to as the *critical range.* In the hardening of steel by heating and quenching, the steel is heated to a definite temperature above that of the critical range.

Softening. After hardening has taken place, either through work hardening or heat treatment, the metal can again be softened. The term softening similarly implies a decrease in the ultimate strength and brittleness (increase in the toughness) as well as a decrease in the hardness. In the case of iron-base alloys, the metal is reheated to a definite point again above the critical range, maintained at this elevated temperature for a definite time, and then cooled very slowly. The name *annealing* is given to this softening process. For most nonferrous alloys, the cooling can be done either slowly or rapidly as by quenching. Annealing also has the effect of relieving the material of stresses caused by cold working (*residual stresses*) and of reorganizing the grain structure back to a normal state. When done with this purpose in mind on iron-base alloys, the process is called *normalizing*. The cooling in normalizing is accomplished in still air.

Tempering. Most of us, at some time or other, have tried to use a knife that was hardened to such an extent that it was too brittle to keep from becoming nicked. The process of *tempering* (also called *drawing back*) aims to eliminate this fault by reintroducing a greater degree of toughness to the hardened steel. Before the steel has cooled to room temperature from the hardening process, it is again heated, this time to a definite temperature *below* the critical range. This temperature is maintained for a period of time, depending on the composition of the steel, and then the steel is allowed to cool either slowly or rapidly by quenching.

Casehardening. Parts which slide or roll on other parts should have hard contact surfaces in order to minimize wear. Yet a hard material throughout for most steels means brittleness and the danger of fracture. Casehardening supplies the answer. In this process a relatively soft, tough steel can be given a hard surface, penetrating to a depth of about $\frac{1}{32}$ in. One method of casehardening is called *carburizing*. The steel is heated to a definite temperature above the critical range for several hours in the presence of a gas, liquid, or solid containing carbon. During this time some carbon is absorbed by the steel in the form of iron carbide, an exceedingly hard compound. Then the steel is heated to a still higher temperature and quenched. The iron carbide is retained near the surface in what is known as a solid solution in the iron. The molecules of each remain mixed.

15. Cast Iron and Wrought Iron. Cast iron, an alloy of iron, carbon, and silicon, is a strong, hard, and brittle metal. The carbon content is always more than 1.7 per cent and often around 3 per cent. The silicon content also varies but in many cast irons is around 2 per cent. As the name implies, cast iron is formed by casting, usually in molds of sand. There are three types, namely, *gray*, *white*, and *malleable* cast iron.

In gray cast iron, some of the carbon is in the form of iron carbide and the remainder is in the free state as graphite, which can be seen as flakes under a microscope. The graphite has a weakening effect and gives the fracture a grayish appearance, thus the reason for the name. Silicon helps to graphitize the carbon during solidification. Gray cast iron is sufficiently soft to be machined.

With low carbon and silicon percentages, white cast iron will be produced on solidification. In this type, all the carbon is in the form of iron carbide to make white cast iron harder, stronger, and more brittle than gray cast iron. The iron carbide gives the fracture a whitish appearance. White cast iron is too hard to be machined.

Cast iron responds to heat treatment in a manner similar to steel. Heating and slow cooling induces softness (annealing), and rapid cooling increases the hardness. In fact, if the percentage of carbon and silicon in the molten metal are such as to produce the softer gray cast iron at a normal rate of cooling, the result may be changed to the harder white cast iron by speeding up the cooling process. Sometimes all or part of the mold is chilled to give the surfaces of the casting extra hardness.

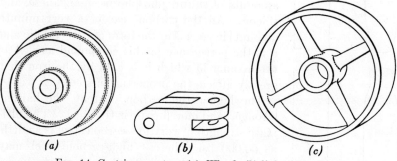

FIG. 14. Cast-iron parts.   (a) Wheel; (b) link; (c) pulley.

Because of its brittleness, cast iron cannot be used where great resistance to shock is important. However, through a process of annealing, a special grade of white cast iron can be *malleablized* to give it more softness and toughness than either the gray or white types. This so-called *malleable iron* can be bent without fracturing immediately after removal from the annealing furnace.

Cast iron finds many uses in industry. Its good wearing quality recommends it for gears and other machine parts. Chilled castings are used to some extent as wheels of railroad cars and malleable cast iron is used in plows, tractors, and various automobile parts (see Fig. 14).

In contrast to cast iron, wrought iron is comparatively weak and soft but of very great toughness. These changes in the properties are accounted for by the almost total absence of carbon in wrought iron. In fact, this metal is an almost pure form of iron. Chief among the impurities (called slag) is iron silicate, which becomes streaked throughout the iron and gives the fracture a fibrous appearance.

Once of great importance, wrought iron today has been almost entirely supplanted by steel. Because it can be hand-forged readily it is used as ornamental iron and blacksmith's work in general. Its great resistance to corrosion and to shock recommends it for certain specialized uses.

16. Steel. Steel, an alloy of iron and carbon up to 1.7, per cent, is the most important of all the metals of industry. Unlike cast iron, which has hardness and strength but not toughness, or wrought iron, which has toughness but not hardness and strength, steel combines in a single material all three of these desirable properties in varying degrees. Plain carbon steel contains around 98 to 99 per cent iron, from 0.05 to 1.7 per cent carbon, and small controlled amounts of sulfur, phosphorus, manganese, and silicon. All the carbon (except a very minute amount) in steel is in the form of iron carbide and it is the percentage of this compound as well as the manner in which it is held by the iron that greatly affects the properties of steel.

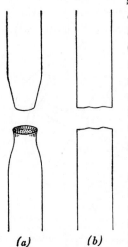

(a)   *(b)*

Fig. 15. Types of tension fractures. (*a*) Mild steel; (*b*) hard steel.

The higher the carbon content, the harder, stronger, and more brittle is the steel. The ultimate strength of very low-carbon steel is as little as 50,000 psi, whereas high-carbon steel may attain a strength of 150,000 psi without special hardening by heat treatment. Plain carbon steel can be classified for convenience into mild (carbon less than 0.3 per cent), medium (carbon from 0.3 to 0.6 per cent), and hard (carbon from 0.6 to 1.7 per cent). Figure 15 shows the fracture in tension of low- and high-carbon test specimens. Note the characteristic *necking down* and the so-called *cup-and-cone fracture* of the mild steel in contrast to the plane fracture of the hard steel.

Heating and quenching, also heating and slow cooling, modify the properties of medium and hard steel by affecting the manner in which the iron carbide is held by the iron. The tensile strength of hard plain carbon steel can be raised by heating and quenching to over 200,000 psi with corresponding increases in hardness and brittleness. In the "as rolled" condition, hard steel must first be annealed in order to be machined, after which it can be hardened to a still harder state than originally. Tempering reintroduces a degree of softness and toughness.

For the same carbon content, cold rolling results in a harder and stronger steel than the hot-rolling process, since the cold-rolled steel is work-hardened. However, hot-rolled steel has greater toughness and ductility.

The mild steels cannot be hardened by heating and quenching, but they do respond to casehardening, which has the advantage of forming a hard exterior for wearing surfaces and retaining the original toughness of the mild steel interior.

One or more elements, such as chromium, vanadium, nickel, or tungsten, are added to straight carbon steel to form *alloy* steels. Some of these so-called alloying elements are held by the iron in solid solution in their element form; others first combine with the carbon. Alloy steels can be classified into two general groups: (1) those used for structural members or machine parts and (2) those designed for specialized uses.

In the first group, the percentages of the alloying elements are small (mostly less than 4 per cent). The purpose in general is to increase the depth to which the steel can be hardened (hardenability) by heating and quenching as well as to keep the toughness in the hardened steel. These steels possess the good qualities of great strength, hardness, and toughness.

TABLE 2

Types of steel	Outstanding properties	Uses
Plain carbon:		
Mild, up to 0.3% C.......	Toughness and less strength	Chains, rivets, nails, structural steel, crane hooks
Medium, 0.3 to 0.6% C....	Toughness and strength	Heat-treated machine parts, railway-car axles, locomotive tires, some tool steel
Hard, 0.6 to 1.7% C......	Less toughness: more hardness	Saws, drills, knives, razors, finishing tools
Nickel..................	Toughness and strength	Rock drills, crankshafts, some structural steel
Chromium-vanadium.......	Hardness and great strength	Automobile gears, propeller shafts, connecting rods
Silicon-manganese..........	Springiness (resilience)	Automobile and railway-car springs
Chromium-nickel (stainless).	Rust resistance	Exterior trim of buildings, surgical instruments, kitchen utensils

In the second group, the percentages of the alloying elements are often much larger (around 18 per cent of chromium in some stainless steels). The purpose here is to impart to the steel one or more special properties. For example, chromium and nickel are used for rust resistance, molybdenum and chromium prevent deterioration of certain properties at high temperature, and cobalt improves the magnetic qualities.

The American Iron and Steel Institute (AISI) has adopted a code by which the process of manufacture, the type of alloy, and the approximate percentage of carbon is revealed by a letter and number. This system is an adaptation of and supersedes the SAE numbering system.

Steel can truly be said to be an indispensable metal in our modern world. Table 2 gives a rough classification of plain carbon steels and some alloy steels, their outstanding properties, and a few of their many uses.

17. Copper. In the world of industry, the element copper ranks next to iron and carbon in importance, chiefly because of its use when alloyed with other elements. Copper is a soft, tough, and ductile metal, and can withstand severe bending and forging without failure. It can be hammered into thin sheets and drawn into wires.

There is no way of heat-treating copper to harden it, but it can be made considerably more hard and strong, but less tough by cold working. In the cast state, copper has a tensile strength of about 25,000 psi, but when work-hardened it reaches a much higher value. By heating and cooling (slow or rapid), copper can be annealed to restore its original properties after having been work-hardened.

Copper is too weak and soft to be used to any extent in machine parts. Owing to its high electric conductivity, it is used extensively in the electrical industry as wires, bus bars, etc. It is rolled into sheets from which tanks and low-pressure boilers are made for special uses in chemical plants. Copper is also used extensively for tubing.

18. Brass. This alloy consists of a mixture of copper and zinc in proportions of 5 to 45 per cent zinc. Because of the complicated way in which zinc dissolves in copper, the varying percentages of the two alloying elements greatly affect the properties of the brass. When the zinc content is below 20 per cent, the product is known as *red brass*— a readily workable, tough, and ductile metal. As the zinc content increases to about 30 per cent, the brass becomes more ductile and (contrary to what you would expect) of greater tensile strength (see Fig. 16). Proportions of between 28 and 35 per cent result in *cartridge brass*, the most ductile of all brasses. As the name indicates, this type of brass is used for stamping and deep drawing, as in the manufacture

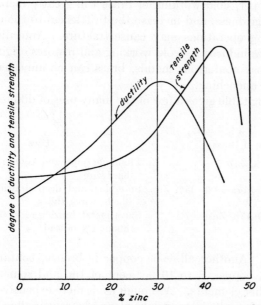

FIG. 16. How the zinc content in brass affects strength and ductility.

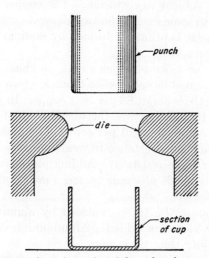

FIG. 17. One of a series of deep-drawing steps.

of cartridge shells (see Fig. 17). A brass of still greater tensile strength, but of less ductility, is made with 40 to 45 per cent zinc, and is known as *Muntz metal*. This product lends itself more to hot working than cold working.

As with copper, cold working of brass will greatly increase tensile strength and hardness, and increase the brittleness to the extent that further working operations may cause fracture. Annealing relieves residual stresses induced by cold working and restores original properties. Like most nonferrous metals, brass can be annealed by rapid cooling—even quenching.

The following table gives a few of the many uses of different types of brass.

Type of Brass	Uses
Red brass (5–20% Zn)	Plumbing pipe and connections, rivets, hardware
Cartridge (yellow) brass (28–35% Zn)	Stamping and deep drawing, cartridge shells, wires, tubes
Muntz metal (40–45% Zn)	Sheet metal, brazing rods, castings (with about 1% of lead)

19. Bronze. Another alloy of copper is bronze, containing 90 to 95 per cent of copper, 5 to 10 per cent of tin, and sometimes small amounts of other elements. Although tin is more expensive than zinc, its use in bronze results in a much harder and stronger alloy than brass. Also bronze is less subject to corrosion. The copper holds the tin in solid solution and to some extent the two elements enter into chemical combination. Bronze is primarily formed by casting, whereas brass is usually formed by working.

Introducing one or more other elements, to change the properties somewhat, results in metals called *alloy bronzes*. *Gun metal*, a metal of considerable strength, is about 88 per cent copper, 10 per cent tin, and 2 per cent zinc. In *nickel bronze* the nickel content runs from very low percentages to more than 10 per cent. The nickel, among other things, serves to increase resistance to wear (hardness) and makes the alloy still more corrosive resistant. Adding phosphorus to bronze in very small amounts yields *phosphor bronze*, a metal of improved resilience and hardness.

In *aluminum bronze* the tin is replaced by aluminum with small amounts of other elements added. Aluminum bronze has greater strength and ductility than tin bronze.

Many of the alloy bronzes, including aluminum bronze, can be heat-treated to improve their properties, particularly to raise the ultimate strength. This process involves heating and quenching.

Bronze, because of its high electric conductivity, finds many uses in the electrical industry. Also, because of its resistance to corrosion, it is used extensively in marine work. Alloy bronzes are used in machines for bushings (linings of bearings) and for gears (mostly nickel bronze). Many springs are made of phosphor bronze.

20. Aluminum and Its Alloys. Aluminum is a soft, weak metal with excellent properties of ductility and malleability. It can be drawn into wires and rolled into thin sheets. Owing to its lack of hardness and strength, however, its use in industry in the element or near-element form is limited.

Various elements are selected to produce aluminum alloys which have much improved properties of hardness and strength. Some of these alloys are particularly adapted to casting; others are adapted to either hot or cold working. For many of the alloys about 4 per cent of copper and about $1\frac{1}{2}$ per cent of silicon are mixed with aluminum. In high-silicon alloys, the silicon may be as much as 12 per cent. For sand castings, zinc in larger percentages generally replaces the silicon. The wrought (suitable for hot or cold working) aluminum alloys usually have smaller percentages of alloying elements.

There are several high-strength aluminum alloys, of which the oldest and most extensively used is *duralumin*. This alloy contains 4 per cent copper with manganese, silicon, and iron totaling 1 per cent. Alloys of this type are successfully used as substitutes for steel.

Both aluminum and its alloys become hardened from cold working and sometimes this effect is much desired in the finished product. As in the case of

Fig. 18. An aluminum-alloy die-cast piston.

many of the alloys previously discussed, aluminum and its alloys can be annealed. Also, similar to steel, the ultimate strength of some of these aluminum alloys can be raised by heating, maintaining the raised temperature over a period of time, and then cooling rapidly by quenching.

Lightness, coupled with other properties necessary in machine parts, has won for aluminum alloys a place wherever excess weight is a problem. In transportation, particularly in the airplane industry, these alloys are used extensively for body and engine parts (Fig. 18).

21. Miscellaneous Alloys. Another group of lightweight alloys are important in industry, namely, the magnesium alloys, known commonly as *Dow metal*. The compositions are magnesium 87 to 95 per cent, aluminum 4 to 12 per cent, and manganese up to about 1.5 per cent. Dow metal is not as strong as some of the aluminum alloys, but it is lighter, and its use is recommended where lightness is of prime importance.

Monel metal stands out among alloys as a natural alloy, by which is meant that it is refined from ores bearing the elements of the finished product, in this case, nickel and copper with traces of iron, manganese, and cobalt.   Nickel comprises about 70 per cent and copper about 29 per cent.   This metal has properties similar to steel (ultimate tensile strength of 80,000 psi) and in addition does not corrode.   It is, therefore, successfully used as a steel substitute.   Its costliness compared to stainless steel now prohibits more general use.

The use of bronze as a bearing metal has already been mentioned. For this use *babbitt metal* is of still greater importance.   The name babbitt metal is applied to the alloys of tin.   Although the amounts of the alloying elements vary, an 85 per cent tin, 10 per cent antimony, and 5 per cent copper combination is common.   Copper acts as a hardener and antimony as a toughener for the soft tin, and, in addition, the tin combines to some extent both with the copper and the antimony to add to the hardness.

PROBLEMS

1. Why are metals in their pure form unsuitable for industrial uses?
2. Name three ways by which the properties of many alloys can be modified.
3. In what way are some of the other properties of an alloy usually affected when it is hardened?
4. Explain what is meant by the critical range.
5. How does normalizing differ from annealing as to purpose and procedure?
6. What is the value of the tempering process?
7. How does the temperature to which the piece of steel is raised in tempering differ from that for hardening or annealing?
8. How does carbon content affect cast iron, wrought iron, and steel with reference to hardness and toughness?
9. What is malleable iron?   What advantages has this form over white or gray cast iron?
10. Some steels must be heat-treated before machining.   Explain.
11. What heat-treatment can be recommended for a low-carbon steel to improve its wearing qualities?
12. What is the chief reason for the use of alloy steels in machine parts?
13. Name one important respect in which the annealing process for brass may differ from that for steel.
14. List the advantages and disadvantages of bronze over brass for industrial uses.
15. Why has aluminum bronze to some extent replaced tin bronze?
16. Why is duralumin said to be a successful steel substitute?
17. What are the reasons that Monel metal is not used more extensively today?

CHAPTER 4

GENERAL PROBLEMS OF FORCE AND MOTION

22. Forces—Their Composition and Resolution. Designing engineers frequently encounter problems in which the forces involved act at various angles to the axis of the member under consideration. Examples are found in the crane, the toggle joint, and the piston and connecting rods in Fig. 19a, b, and c. To what extent are the various members stressed in each case when resisting the external force P?

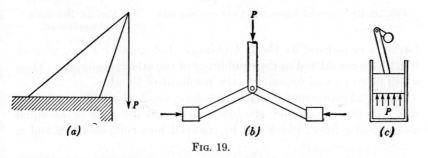

(a)   *(b)*   *(c)*

Fig. 19.

To answer this question, we shall begin by reviewing briefly a part of physics. In Fig. 20a forces a and b are *concurrent*, which means that their lines of action meet at a point. If these two forces are replaced by force c of such a magnitude that c has the same effect as a and b, then c is the *resultant* of a and b. Similarly in Fig. 20b if the force c is replaced by concurrent forces a and b both of which have the same effect as c, then a and b are *components* of c. The process of determining the resultant is called *composition*, whereas that of determining the components is called *resolution*.

In Fig. 21 the three concurrent forces (a, b, and c) cause no motion. When this condition obtains, the forces are said to be in equilibrium. Any one of the three forces is needed to hold the other two in equilibrium, and, hence, is called the *equilibrant* of the other two. If any two of the forces, as a and b, are replaced by their resultant d, equi-

27

librium will still exist, since d has the same effect as a and b. It follows that d and c must be equal in magnitude but opposite in direction; otherwise motion would result.

The relationship between the resultant of two or more forces and their equilibrant always holds true no matter which of the original

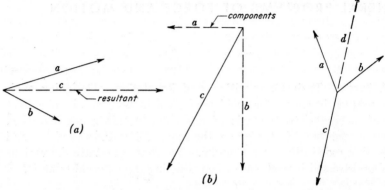

(a)

(b)

FIG. 20. Resultant of forces and force components.   FIG. 21. Resultant and equilibrant.

forces is considered as the equilibrant. Let another force, as a of Fig. 21, be considered as the equilibrant of the other two forces. Then a will be equal and opposite to the resultant of b and c.

Figure 22 represents a member pushing on the base at an angle of 60°. To prevent motion, the base must resist the push by an equal and opposite force, which can be resolved into components q and r.

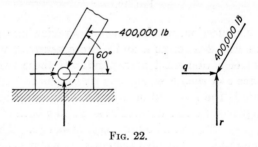

FIG. 22.

Hence, concurrent forces 400,000 lb, q, and r are in equilibrium. What are the values of q and r?

23. Graphical Solutions. The values of q and r can be found both graphically and analytically. Although graphical solutions vary somewhat in detail, they all adhere to the following general pattern, as represented in Fig. 23a. Draw an arrow for the 400,000-lb force parallel to the direction of this force, that is, 60° to the horizontal. The length of the arrow should be made to a suitable scale to represent 400,000 lb magnitude. When an arrow thus shows magnitude as well as direction, it is known as a *vector*. At the head and tail of the arrow, lines are drawn parallel to q and r to form a closed polygon, called the *force polygon* (in this case a triangle). The lengths of the two sides of the triangle represent the magnitudes of forces q and r.

Often the directions of the various forces are not as obvious as they are in this problem. However, to find their direction, the following rule holds true for any number of forces in equilibrium. Select the sequence of plotting the forces by going around the joint, through which the forces act, either in a clockwise or counterclockwise direction. Then plot the vectors in the sequence selected with one beginning at the point where the other ends. This diagram should be a closed polygon for forces in equilibrium. The direction of one or more of the forces in the polygon is known and the directions of the others follow forward in the same path on the perimeter, with the head of each arrow touching the tail of the adjacent arrow (Fig. 23b). For example, 400,000 lb, r,

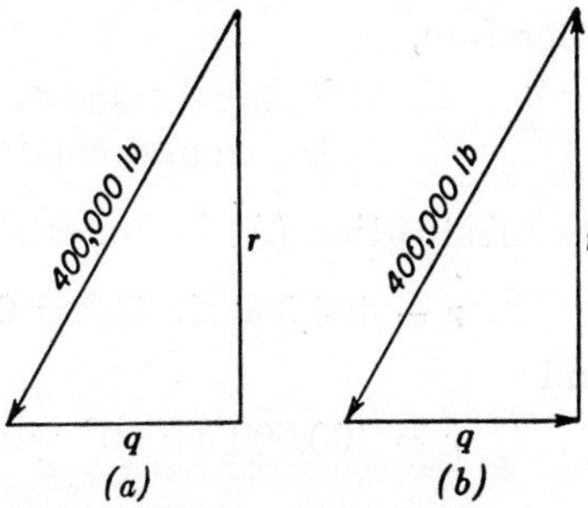

FIG. 23. Graphical solution.

and q form a clockwise sequence of the forces about the joint of Fig. 22, whereas 400,000 lb, q, and r form a counterclockwise sequence. In Fig. 23a the counterclockwise sequence was selected and the polygon started by plotting the vector of 400,000 lb. Then q followed from the point where 400,000 lb ends and, lastly, r was drawn to form the closed polygon. The direction of 400,000 lb (being known) starts us along the perimeter and q and r follow forward in the same path as shown by the arrowheads of Fig. 23b.

24. Analytical Solutions. Both graphical and analytical solutions of this and similar problems are based on the principle of physics that for equilibrium to exist the algebraic sum of the vertical components of all the forces must equal zero and the algebraic sum of the horizontal components of all the forces must also equal zero, abbreviated $\Sigma V = 0$ and $\Sigma H = 0$. Since the lines of action of all the forces of Fig. 22 meet at a point, there is no tendency toward rotation and, hence, moments need not be considered.

In the analytical solution, upward forces are usually considered as plus and downward forces as minus. Also forces acting to the right are plus, while those to the left are minus. With this in mind and by the use of the principle that $\Sigma V = 0$ and $\Sigma H = 0$, we can form the following equations for this problem:

$$r - \text{vertical component of } 400{,}000 = 0 \qquad \text{(A)}$$
$$q - \text{horizontal component of } 400{,}000 = 0 \qquad \text{(B)}$$

Now consider the fact that

$$\sin 60° = \frac{\text{vertical component of } 400{,}000}{400{,}000}$$

and

$$\cos 60° = \frac{\text{horizontal component of } 400{,}000}{400{,}000}$$

Therefore,

$$\text{Vertical component} = 400{,}000 \sin 60°$$
$$\text{Horizontal component} = 400{,}000 \cos 60°$$

By substitution (A) becomes

$$r - 400{,}000 \sin 60° = 0$$

and

$$r = 400{,}000 \sin 60° = 400{,}000 \times 0.866 = 346{,}000 \text{ lb*}$$

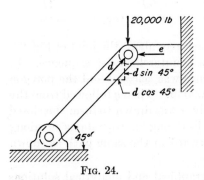

By substitution (B) becomes

$$q - 400{,}000 \cos 60° = 0$$

and

$$q = 400{,}000 \cos 60°$$
$$= 400{,}000 \times 0.5 = 200{,}000 \text{ lb}$$

Fig. 24.

Become accustomed to thinking of the two smaller sides of the right triangle as the product of the hypotenuse and the sine or cosine of an angle. In future illustrative problems the final form of (A) and (B) as given above will be set down directly.

The total stresses d and e in the members of Fig. 24 owing to the force of 20,000 lb acting vertically on the joint are again determined by $\Sigma V = 0$ and $\Sigma H = 0$.

*Since slide-rule work is sufficiently accurate for problems relating to machine design, use of the sliderule is preferred. Calculations in this text were so made. Slight discrepancies in numerical answers may be expected.

$\Sigma V = 0$:
$$d \sin 45° - 20{,}000 = 0 \tag{A}$$

$\Sigma H = 0$:
$$d \cos 45° - e = 0 \tag{B}$$

From (A),
$$d \times 0.707 - 20{,}000 = 0$$
$$d = \frac{20{,}000}{0.707} = 28{,}300 \text{ lb}$$

Substituting this value of d in (B),
$$28{,}300 \times 0.707 - e = 0$$
$$e = 20{,}000 \text{ lb}$$

The weight suspended from the cables in Fig. 25a causes tension c and k in the left and right cables, respectively. Forces c and k must hold the weight in balance; otherwise motion would result. The

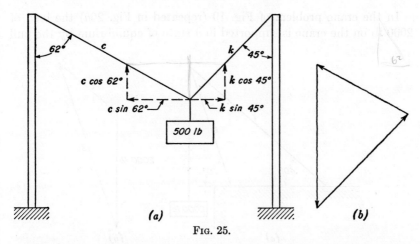

$$62°\quad c$$
$$k\quad 45°$$
$$c \cos 62°\qquad\qquad k \cos 45°$$
$$c \sin 62°\qquad\qquad k \sin 45°$$
$$\boxed{500\ lb}$$

(a)   (b)

Fig. 25.

weight is the equilibrant of c and k, but it is also true that *each* force can be considered as the equilibrant of the other two. A rough sketch of the graphical solution in Fig. 25b is useful in verifying the directions of the forces. The vertical and horizontal components of c and k are shown in Fig. 25a. The two equations for $\Sigma V = 0$ and $\Sigma H = 0$ are as follows:

$\Sigma V = 0$:
$$c \cos 62° + k \cos 45° - 500 = 0 \tag{A}$$

$\Sigma H = 0$:
$$- c \sin 62° + k \sin 45° = 0 \tag{B}$$

From (B),

$$c = \frac{k \sin 45°}{\sin 62°}$$

Substituting in (A),

$$\frac{k \sin 45° \cos 62°}{\sin 62°} + k \cos 45° - 500 = 0$$

$$\frac{k \times 0.707 \times 0.470}{0.883} + 0.707k - 500 = 0$$

$$0.376k + 0.707k - 500 = 0$$

$$1.083k = 500$$

$$k = \frac{500}{1.083} = 461 \text{ lb}$$

Substituting in (B),

$$c = \frac{461 \times 0.707}{0.883} = 370 \text{ lb}$$

In the crane problem of Fig. 19 (repeated in Fig. 26a) the load of 2000 lb on the crane is supported in a state of equilibrium by the pull

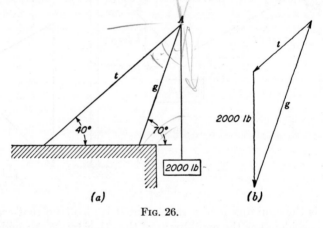

(a)   (b)

FIG. 26.

or push of t and g. Hence, this load is the equilibrant of t and g. A rough sketch of the graphical solution (Fig. 26b) shows the direction of the stresses t and g. Since g pushes toward the joint A, g causes a compressive stress, and since t pulls away from the joint, t causes a tensile stress, in the respective members.

Again we can form equations to calculate analytically the values of the total stresses g and t.

$\Sigma V = 0$:

$$g \sin 70° - t \sin 40° - 2000 = 0 \qquad \text{(A)}$$

$\Sigma H = 0$:

$$g \cos 70° - t \cos 40° = 0 \qquad (B)$$

From (B),

$$g = \frac{t \cos 40°}{\cos 70°}$$

Substituting in (A),

$$\frac{t \cos 40° \sin 70°}{\cos 70°} - t \sin 40° - 2000 = 0$$

$$\frac{t \times 0.766 \times 0.940}{0.342} - t \times 0.643 - 2000 = 0$$

$$2.110t - 0.643t - 2000 = 0$$
$$1.467t = 2000$$
$$t = \frac{2000}{1.467} = 1360 \text{ lb}$$

Substituting in (B),

$$g = \frac{t \cos 40°}{\cos 70°} = \frac{1360 \times 0.766}{0.342} = 3050 \text{ lb}$$

25. Force, Motion, and Equilibrium. In the problems so far discussed, as well as those encountered for the most part in strength of materials, the forces involved were considered as static, that is, they caused no motion. As a matter of fact in designing members under static loading conditions, motion was usually just the thing to be avoided. Motion signified breaking (failure) and showed that the design was faulty. The three conditions of equilibrium are:

1. The algebraic sum of the upward and downward components of all the forces must equal zero ($\Sigma V = 0$).
2. The algebraic sum of the components to the left and to the right of all the forces must equal zero ($\Sigma H = 0$).
3. The algebraic sum of the moments about any point as a center of moments must equal zero ($\Sigma M = 0$).

These conditions were applied to bodies at rest, and if these conditions were satisfied, the forces and moments were in balance and equilibrium obtained. When motion resulted, the body was no longer in equilibrium.

The mechanical or electrical engineer, on the other hand, is particularly interested in constructing objects with moving parts, such as machines, engines, motors, and generators. To him equilibrium again means balanced forces and moments, but more often than not the body on which the forces act is in motion.

How is it possible to have equilibrium and motion at the same time? To answer this question, let us consider an example of motion where a locomotive is pulling a train of cars (Fig. 27). Assume that the force of the locomotive on the cars is just equal to the frictional force of the cars tending to hold back the train. As long as this condition exists, one force balances the other and the speed of the train does not change. To speed up, the pulling force must be increased; and to slow down, the pulling force is decreased. When the forces are unbalanced, the train will either continue to speed up or come to a stop. The engine of the locomotive (and all machines and engines for that matter) act similarly. At start the moment causing an engine or motor to rotate overbalances the moment resisting such rotation, when rotating at constant speed the moments are in balance, and before stopping the moments are again unbalanced, this time the moment resisting rotation is the greater. Since we must design for maximum stress con-

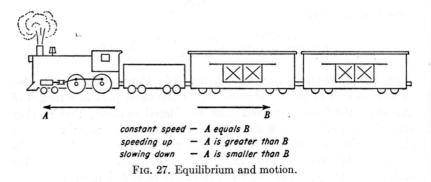

constant speed — A equals B
speeding up   — A is greater than B
slowing down   — A is smaller than B

Fig. 27. Equilibrium and motion.

ditions, the forces or moments causing the maximum stress are used in design calculations. In Fig. 27 the coupling between the locomotive and the tender must resist the greatest force when the train is picking up speed at the maximum rate.

The crane problem of the previous section offers an example of force causing motion. When the crane holds the body that it has lifted suspended in a state of rest, the force that the crane must support is just equal to the weight of the body. This condition again prevails when the body is being lifted or lowered at a constant rate of speed. But when the crane is lifting the body at an accelerated speed (speeding up), the required force is greater because the crane must not only overcome the force owing to the attraction of gravity but must supply sufficient additional force to cause acceleration in the upward direction. This condition results in the greatest stress in the crane members supporting the weight and is assumed in designing these members.

To determine the value of the force on the cable during periods of acceleration, we turn to Newton's second law (law of acceleration).

Newton discovered that the acceleration of a body varies as the force causing that acceleration—the greater the force, the greater the acceleration. It follows that, if two forces are acting on a body, the accelerations a and a_1 that are produced vary as do the magnitudes of the forces F and F_1. In symbols,

$$\frac{F}{F_1} = \frac{a}{a_1}$$

One force always acting on a body is the force of gravity which is equal to the body's weight W. Also we know that the acceleration produced by gravity, g, is 32 ft per sec per sec. Hence, the above equation can be written,

$$\frac{F}{W} = \frac{a}{g} \quad \text{and also} \quad F = \frac{Wa}{g} \tag{4}$$

Therefore, when the weight of the body to be hoisted is known and the acceleration that can be produced on that body is known, then the

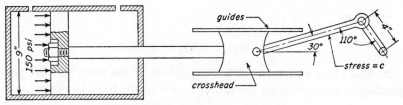

FIG. 28. Reciprocating steam engine.

additional force F on the cable of the crane needed to produce that acceleration can be calculated. To find the total force on the cable, the two forces are added.

$$\text{Total force} = W + F$$

This total force must be used to calculate the maximum stress in the framework of the crane.

26. Problems of Forces in Motion. The steam-engine cylinder of Fig. 28 affords another illustration of forces causing motion. As long as the mechanism attached to the piston rod and the friction of the piston offer a resistance equal to or greater than the force of the steam on the piston, no motion will take place. Motion starts at the instant the piston force becomes greater than this resistance and speed increases while this condition prevails. During a period of uniform motion the piston force and the resistance are again equal and the system is once more in equilibrium.

The problem is to determine (1) the total stress c in the connecting rod; (2) the side thrust on the crosshead; (3) the value of the twisting moment (torque) of the crankshaft.

1. The push of the piston is horizontal, whereas the connecting rod in the position indicated pushes at an angle of 30° to the horizontal. However, the horizontal component of the push c of the connecting rod must equal the push of the piston.

Area of piston $= 0.785 D^2 = 0.785(9)^2 = 63.6$ sq in.

Force of piston $=$ area $\times$ pressure $= 63.6 \times 150 = 9540$ lb

$\Sigma H = 0$:

$$9540 - c \cos 30° = 0$$
$$9540 - c \times 0.866 = 0$$
$$c = \frac{9540}{0.866} = 11,000 \text{ lb}$$

2. The part of the force c that acts vertically transmits a side thrust h to the crosshead. Therefore, since

$\Sigma V = 0$:

$$h = c \sin 30°$$
$$h = 11,000 \times 0.500 = 5,500 \text{ lb}$$

3. The value of a moment is the product of the force and the perpendicular distance from the line of action of the force to the point of

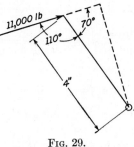

Fig. 29.

rotation. In this case the given distance of 4 in. is measured along a line making an angle of 70° with the line of action of the force as illustrated in Fig. 29. Hence $4 \times \sin 70°$ is the perpendicular distance, and

Torque $= 11,000 \times 4 \sin 70° = 11,000 \times 4 \times 0.940 = 41,400$ lb-in.

Another problem of force and motion is that of the toggle joint of Fig. 30a in which we are to calculate the length of m and n necessary to create vertical forces of 120 lb each at the ends of these members when the joint is 2 in. distant from the center line QQ. Figure 30b shows this relationship for member m. If angle α were known, the length of m could be calculated, for

$$\cos \alpha = \frac{2}{m}$$

and

$$m = \frac{2}{\cos \alpha}$$

On the other hand, referring to the force diagram of Fig. 30c, we note that

$$\tan \alpha = \frac{\text{vertical component of force } m}{\text{horizontal component of force } m}$$

and both components are known. The vertical component is 120 lb and, since both m and n are pushing on the joint equally, the hori-

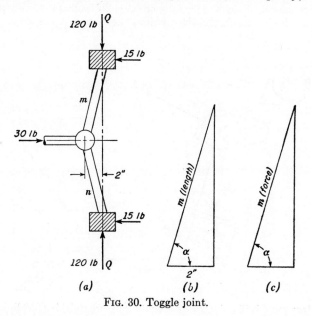

FIG. 30. Toggle joint.

zontal component of each is 15 lb acting to the left to balance the push of 30 lb acting to the right. Therefore,

$$\tan \alpha = \frac{120}{15} = 8$$

$$\alpha = 82°\text{-}52'$$

and again,

$$m = \frac{2}{\cos \alpha} = \frac{2}{0.124} = 16.1 \text{ in.}$$

where m is the length.

PROBLEMS

1. A canalboat is towed by a pull of 440 lb at an angle of 12° with the axis of the boat. Calculate the force applied to the boat:

    a. In the line of its axis

    b. At right angles to its axis

2. A steam pipe is supported by a series of A frames. Each frame holds a length of pipe weighing 1600 lb suspended from the top of the A. Each side is 8 ft long and 6 ft distant from the other at the base. The bottom ends are embedded in concrete. Calculate the total stress:

    a. In each of the side members

    b. In the horizontal crosspiece of the A, assuming that the sides do not bend

3. Because of the obstruction caused by shaft S the brake is applied in the manner here shown. Calculate the force on the cable leading to the brake. *Hint:* Problems of this kind are simplified by rotating the figure to a position in which one or more forces are either vertical or horizontal.

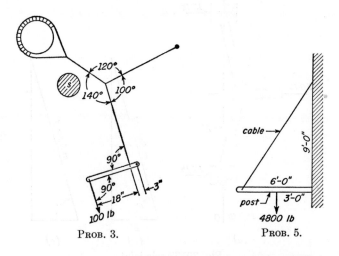

PROB. 3.   PROB. 5.

4. A large room is illuminated by a lighting fixture weighing 350 lb hung from the ceiling by a chain 7 ft 0 in. long. For purposes of painting the ceiling the fixture is pulled horizontally a distance of 2 ft by means of a wire fastened at the end of the chain. Calculate the tension in the wire.

5. *a.* Calculate the tension in the cable and the compression in the post of the jib crane here shown. Neglect the bending stress developed in the post.

    b. Calculate the required cross-sectional area of the cable if a stress of 12,000 psi is not to be exceeded.

6. A structural beam 10 ft long and 50 lb per ft in weight is hung from a crane hook by means of cables fastened to the ends of the beam. The angle between cables is 30°. Calculate the tension in the cables.

7. The force F is used to hold the weight of 6000 lb in the accompanying figure in a state of equilibrium. The pulley is assumed to be frictionless. Calculate the total compressive stress:

    a. In member AB
    b. In member CB

Hint: The tension in all portions of a rope over a frictionless pulley is constant. Hence, F equals the weight of 6000 lb.

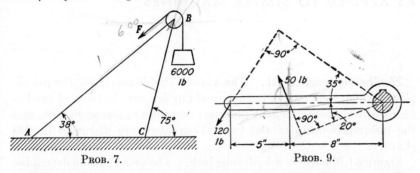

PROB. 7.　　　　　　PROB. 9.

8. An elevator car plus its capacity load of passengers weighs 4 tons. The maximum acceleration of the car is 10 ft per sec per sec in an upward direction. Calculate the total pull on the cables.

9. Calculate the net torque on the shaft caused by the forces on the lever in the figure here shown.

10. Calculate the torque developed on the crankshaft by the gasoline engine of the figure when the angle α is a maximum. Assume a pressure of 300 psi on the piston at this point in the power stroke.

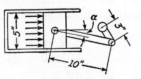

PROB. 10.

11. A toggle-joint mechanism similar to that of Fig. 30 is acted on by a horizontal force of 500 lb at the joint to create vertical forces of 3600 lb at the ends of the arms. If each arm is 38 in. in length, calculate:

    a. The distance from the joint to the line of action of the 3600-lb forces
    b. The value of the vertical forces when the joint is 5 in. away from their line of action.

CHAPTER 5

FORCE AND MOTION
AS APPLIED TO SIMPLE MACHINES

27. The Hydraulic Jack. The various simple machines, the principles of which were part of our study of physics, are the bases of mechanisms which afford excellent practical examples of forces, motion, and the interrelationship of the two. Four cases are discussed in this chapter.

Figure 31 illustrates a hydraulic jack. The problem is to determine (1) the lifting capacity of the jack when a force of 25 lb is applied to the end of the lever (friction neglected); (2) the pressure in pounds per square inch in the large cylinder.

1. As a step in the solution of this part, the force of the piston resisting the push of the handle can be calculated. This force (which we shall denote as R) has been increased from the original force of 25 lb by means of the lever. To determine its value, we take moments about the fulcrum, giving the clockwise moments a plus sign and the counterclockwise moments a minus sign.

$$-\tfrac{3}{2} \times R + 25 \times 10 = 0$$
$$\tfrac{3}{2} \times R = 25 \times 10$$
$$R = \frac{25 \times 10 \times 2}{3} = 167 \text{ lb}$$

The forces of the pistons are in direct proportion to their areas. In this case

$$\frac{P}{167} = \frac{0.785(\tfrac{3}{2})^2}{0.785(\tfrac{3}{4})^2}$$

where P is the large piston force or the lifting power of the jack. By canceling out 0.785,

$$\frac{P}{167} = \frac{(\tfrac{3}{2})^2}{(\tfrac{3}{4})^2}$$

and

$$P = \tfrac{3}{2} \times \tfrac{3}{2} \times \tfrac{4}{3} \times \tfrac{4}{3} \times 167 = 668 \text{ lb}$$

2. According to Pascal's principle, the pressure p (in pounds per square inch) in the large cylinder, required in part 2, is the same as the pressure at any other point in the interior of the jack. Its value

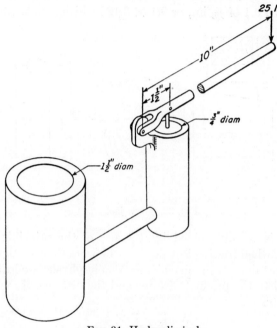

FIG. 31. Hydraulic jack.

can, therefore, be determined by dividing either force by the respective cylinder area:

$$p = \frac{668}{0.785(\tfrac{3}{2})^2} = \frac{668 \times 2 \times 2}{0.785 \times 3 \times 3} = 378 \text{ psi}$$

also

$$p = \frac{167}{0.785(\tfrac{3}{4})^2} = \frac{167 \times 4 \times 4}{0.785 \times 3 \times 3} = 378 \text{ psi}$$

28. The Screw Jack. The machine represented in Fig. 32 is in reality a combination of three machines, namely, the handle (lever), the screw (inclined plane), and the wedge (inclined plane). We are to calculate (1) the distance that the 1600-lb weight will be lifted by one turn of the handwheel; (2) the force that must be applied to the handwheel to lift this weight; (3) the theoretical mechanical advantage of the system.

1. Since there are 10 threads to the inch, the pitch of the screw is 0.1 in., and for one turn of the handwheel the shank will advance 0.1 in. horizontally (assuming a single-thread screw). The slope of the wedge lifting the weight is ½:4 or 1:8. This means that the weight is lifted ⅛ in. for a 1-in. horizontal advance of the wedge. For a 0.1-in. horizontal advance (one turn of the handwheel), the weight is lifted 0.1 of ⅛ in., or 0.1 × 0.125 which equals 0.0125 in.

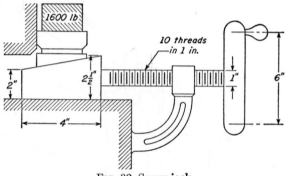

FIG. 32. Screw jack.

2. Part 2 is readily solved by the use of the principle of equal work, that is, when friction is neglected, the work done *by* the machine is equal to that done *to* the machine. Also in physics we learned that work is equal to the product of force and the distance through which that force acts.

$$\text{Work} = F \times s \tag{5}$$

where s is the symbol for distance. In the case of the screw jack, the work done by the machine in raising the weight must (neglecting friction) equal the work done by the operator in turning the handwheel, or

$$F \times s = F' \times s'$$

By making substitutions in the above equation for the values that the weight moves for one turn of the handwheel, we have

$$1600 \times 0.0125 = F' \times \pi \times 6$$

where $\pi \times 6$ is equal to the distance traveled by the handle of the handwheel during one revolution. Then

$$F' = \frac{1600 \times 0.0125}{3.14 \times 6} = 1.06 \text{ lb}$$

We have now found the theoretical force necessary to move the weight.

You know that the actual force, because of friction, would be considerably more.

3. Theoretical mechanical advantage (TMA) is the number of times the machine would multiply the force if there were no friction. In this case

$$\text{TMA} = \frac{1600}{1.06} = 1510$$

The ratio of the distance the handle of the wheel moves (effort distance) to the distance that the weight moves (resistance distance) is another way to calculate theoretical mechanical advantage, that is,

$$\text{TMA} = \frac{3.14 \times 6}{0.0125} = 1510$$

Note the very large value of the mechanical advantage of this compound machine, accounted for by the fact that the mechanical advantage of a compound machine is the product of the mechanical advantages of the various parts.
Check this statement by determining the mechanical advantage of each part, namely, the lever of the wheel, the inclined plane of the screw, and the inclined plane of the wedge, and then multiplying these values together to see if you obtain 1510.

29. The Gear Train. Circles B, C, D, and E of Fig. 33 represent gears, mating at the points of tangency. Hence, rotation of the crank causes drum F to raise or lower the weight. Diameters of gears B, C, and D are 6, 32, and 20 in., respectively. The problem is to calculate (1) the diameter of gear E necessary to maintain the system in balance, when friction is neglected; (2) the force needed to push the crank handle downward, if the efficiency of the system is 75 per cent.

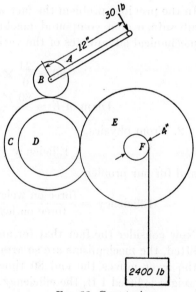

Fig. 33. Gear train.

1. For one revolution of crank A the handle travels in a circular path a distance of $\pi \times 24$ in. Similarly, a point on the circumference of gear B travels $\pi \times 6$ in. The theoretical mechanical advantage of

lever A and gear B taken by themselves as a simple machine is the ratio of these distances.

$$\text{TMA of } A \text{ and } B = \frac{\pi \times 24}{\pi \times 6} = \frac{24}{6}$$

In like manner the mechanical advantage of gears C and D taken by themselves is the ratio of the two diameters; that is,

$$\text{TMA of } C \text{ and } D = \frac{\text{diam of } C}{\text{diam of } D} = \frac{32}{20}$$

and

$$\text{TMA of } E \text{ and } F = \frac{\text{diam of } E}{\text{diam of } F} = \frac{\text{diam of } E}{4}$$

In each case the two gears on one shaft may be considered as a simple machine. Also the mechanical advantage of the system is the ratio of the resistance to the effort.

$$\text{TMA of system} = \frac{\text{resistance}}{\text{effort}} = \frac{2400}{30} = 80$$

In the previous problem the fact was brought out that the mechanical advantage of a compound machine is equal to the product of the mechanical advantages of the various parts. Hence,

$$80 = \frac{24}{6} \times \frac{32}{20} \times \frac{\text{diam of } E}{4}$$

$$\text{Diam of } E = \frac{80 \times 6 \times 20 \times 4}{24 \times 32} = 50 \text{ in.}$$

2. From physics,

$$\text{Efficiency} = \frac{\text{work output}}{\text{work input}}$$

and for our problem,

$$\text{Efficiency} = \frac{\text{force on weight} \times \text{distance weight moves}}{\text{force on lever} \times \text{distance lever moves}}$$

Now consider the fact that for any given distance that the weight is lifted, the mechanisms are so arranged that the force P at the end of the lever moves the end 80 times that distance. Therefore, if the weight is lifted 1 ft, the efficiency equation can be written

$$\text{Efficiency} = \frac{2400 \times 1}{P \times 80}$$

Substituting 75 per cent for the efficiency and transforming,

$$P = \frac{2400}{0.75 \times 80} = 40 \text{ lb}$$

The practical mechanical advantage (PMA), that is, the mechanical advantage when friction is taken into account, is the *actual* number of times the machine multiplies the force.   In this case

$$\text{PMA} = \frac{\text{resistance}}{\text{effort (actual)}} = \frac{2400}{40} = 60$$

Note that the efficiency can be calculated quickly if both **TMA** and **PMA** are known, for

$$\text{Efficiency} = \frac{\text{PMA}}{\text{TMA}} = \frac{60}{80} = 75\%$$

30. The Flywheel and Punch Press.   The cast-iron flywheel shown in Fig. 34a is attached to a punch press.   The wheel weighs 1200 lb and rotates at a speed of 150 rpm.   What is the largest diameter hole that it can punch in a ¾-in. steel plate (Fig. 34b)?   Assume an ultimate strength in shear of the steel plate as 60,000 psi and the weight of cast iron as 0.25 lb per cu in.

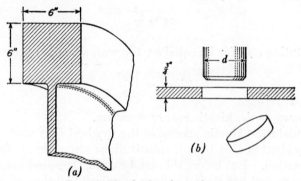

Fig. 34. Flywheel and punch press.

The larger the diameter of the hole, the more work is needed to punch the ¾-in. plate.   The maximum diameter hole is, therefore, limited by the amount of energy supplied by the flywheel.   Assuming that all the energy of the flywheel is absorbed by the punch, we can say

Energy of flywheel = work done by punch

What is the energy of the flywheel?   Energy is the ability to do work, and

$$\text{Work} = F \times s \tag{5}$$

where F = force
s = distance through which force acts

In order to arrive at a value of the force, we again make use of Newton's second law of motion which states that the acceleration of a body varies as the force causing this acceleration. As developed in Sec. 25,

$$F = \frac{Wa}{g} \tag{4}$$

Substituting Wa/g for its equal F in the equation of work as given above, we have

$$\text{Work} = \frac{Wa}{g} \times s$$

If it is assumed that the flywheel slows down (is negatively accelerated) at a constant rate as the punch acts on the plate and finally comes to rest, we can make use of the following equation for constant accelerated motion when starting from or coming to rest:

$$v^2 = 2as \tag{6}$$

where v is the maximum velocity, which in this case is at the start of the punching action. Also

$$as = \frac{v^2}{2}$$

By substituting $v^2/2$ for its equal as, the work equation becomes

$$\text{Work} = \frac{Wv^2}{2g} \tag{7}$$

This is known as the kinetic energy equation.

To calculate the kinetic energy of the flywheel (the work that the moving flywheel is able to do), substitutions are made in the kinetic energy equation. But before this can be done, the speed in rpm must be changed to feet per second and, as practically all the weight of the flywheel is in the rim, the linear speed at the center line of the rim can be taken as the linear speed of the flywheel. First, the circumference of the centerline circle must be determined. This can be done if the volume of the flywheel is known. Since cast iron weighs 0.25 lb per cu in.,

$$W = 0.25 \times V$$

where W = weight of flywheel
 V = volume of flywheel, cu in.

But the volume of the flywheel (flywheel rim) is also

$$V = 6 \times 6 \times C$$

where C is the circumference of the centerline circle of the rim.

Therefore,
$$W = 0.25 \times 6 \times 6 \times C$$
and
$$C = \frac{W}{0.25 \times 6 \times 6} = \frac{1200}{0.25 \times 6 \times 6} = 133 \text{ in.}$$
Also
$$\text{Speed} = \text{rps} \times C = {}^{150}\!\!/_{60} \times {}^{133}\!\!/_{12} = 27.7 \text{ ft per sec}$$

where C is in feet. From the kinetic energy equation,

$$\text{Kinetic energy} = \frac{Wv^2}{2g} = \frac{1200 \times 27.7 \times 27.7}{2 \times 32} = 14{,}400 \text{ ft-lb} \quad (7)$$

As previously indicated, the work done in punching the hole in the steel plate is assumed to be equal to the kinetic energy just calculated. The maximum force needed when the punch initially contacts the plate reduces to zero as the hole is completed. If we call the maximum force P, then $P/2$ is the average force needed, and work done is this force multiplied by the distance through which the force acts ($\frac{3}{4}$ in.).

$$\text{Work done} = \frac{P}{2} \times \frac{3}{4} \times \frac{1}{12} \quad \text{ft-lb}$$

Also
$$14{,}400 = \frac{P}{2} \times \frac{3}{4} \times \frac{1}{12}$$
$$P = 14{,}400 \times 12 \times \frac{4}{3} \times 2 = 461{,}000 \text{ lb}$$

The resisting area to the shearing force is the circumference of the hole multiplied by the thickness t of the plate, and for every square inch of sheared area, the plate offers a resistance equal to its ultimate strength in shear. Therefore,

$$P = \pi d t s$$

where s is the ultimate shearing strength. By substitution,

$$461{,}000 = 3.14 \times d \times \tfrac{3}{4} \times 60{,}000$$
$$d = \frac{461{,}000 \times 4}{3.14 \times 3 \times 60{,}000} = 3.26 \text{ in., or } 3\tfrac{1}{4} \text{ in.}$$

PROBLEMS

1. A hydraulic automobile jack is to be designed to lift 2000 lb. The pump lever is 8 in. long and is connected to the fulcrum at one end. The small plunger has a diameter of $\frac{5}{8}$ in. and is connected to the lever at a distance of $1\frac{1}{4}$ in. from the fulcrum. The maximum effort force is 35 lb. Neglecting frictional losses, calculate:

a. The diameter of the large plunger

b. The pressure in terms of pounds per square inch on the large plunger

2. In the hydraulic part of the pipe bender shown herewith, a force of 2 tons is to be exerted by the piston rod. Specify the minimum length of handle that would operate the small plunger while the operator applies a force of 40 lb on the ball at the end of the handle. Neglect friction.

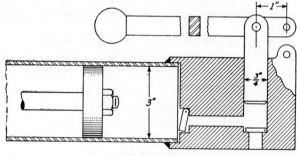

PROB. 2.

3. A load of 36 lb is to be moved up an inclined plane 40 in. long by a force parallel to the plane. The plane makes an angle of 26° with the horizontal. Calculate:

a. The force perpendicular to the plane

b. The force needed to cause upward motion when there is no friction

c. The work done in raising the weight to the top

d. The theoretical mechanical advantage of the machine

4. A screw jack lifts 2200 lb through a distance of 0.065 in. for every turn of the handle. The work lost in friction for each turn is 50 in.-lb. What is the efficiency of the jack?

5. The simple screw jack shown in the accompanying figure raises 1500 lb with a force of 10½ lb at the end of the handle. Calculate:

a. The efficiency of the jack

b. The per cent of work lost in friction

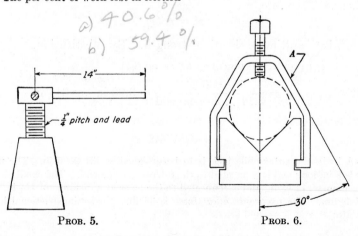

PROB. 5.   PROB. 6.

6. A wrench is used to tighten the clamp here illustrated. A man pulls with a force of 40 lb on the wrench at a distance of 6 in. from the center of the $\frac{3}{8}$–16 ($\frac{3}{8}$ diam, 16 threads to the inch) screw.

a. What force will be exerted on the work by the screw?   Assume a screw efficiency of 25 per cent.

b. What is the stress in the strap at the point *A* if its dimensions are $\frac{3}{8}$ in. by 1 in?

7. *a.* Calculate the theoretical mechanical advantage of the mechanism of Fig. 32 by multiplying the mechanical advantages of the various simple machines represented.

b. Calculate the efficiency of the compound machine, assuming that with the force of 1.06 lb, as calculated for the handwheel, a force of 1050 lb only can be lifted.

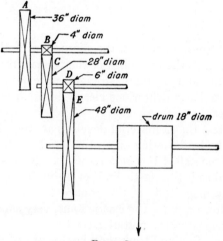

PROB. 8.

a) 75 *lb*
b) 71.4%

8. *a.* The chain pull required on the drum of the power crane mechanism here illustrated is 8400 lb.   Calculate the force that must be applied by an engine to the teeth of gear *A*.   Assume 100 per cent efficiency.

b. Calculate the efficiency of the machine when the chain pull is 4000 lb and the force on the teeth of gear *A* is 50 lb.

9. A weight of 150 lb is moved up an inclined plane a distance of 20 ft by a total force of 115 lb acting parallel to the plane which makes an angle of 30° with the horizontal.   The frictional resistance equals 25 lb.   Calculate the velocity of the weight after the 20-ft run.   *Hint:* For any distance the work done by the *unbalanced* part of the pull equals the kinetic energy at the end of this distance.

11.3 ft/sec

10. A flywheel similar to that of Fig. 34 turns at a velocity of 200 rpm.   The cross-sectional dimensions of the rim are 6 by 8 in. and the circumference of the center line of the rim is 154 in.   Calculate the kinetic energy of the wheel.

11. What is the weight needed in a flywheel rim to lift a load of 400 lb a distance of 10 ft, when the maximum speed is 120 rpm and the diameter of the rim center line circle is 35 in?

12. Holes of 3 in. in diameter are punched in a steel plate, $\frac{1}{2}$ in. thick by a punch press to which a cast-iron flywheel is attached.   The diameter of the rim center-line circle is 48 in. and the speed is 100 rpm.   Calculate the necessary cross-sectional area of the rim.   Assume an ultimate strength of the steel as 60,000 psi and the weight of cast iron as 0.25 lb per cu in.

CHAPTER 6

DESIGN STRESS AND DYNAMIC LOADING

31. Induced Stress. Prior to the discussion of the properties of materials in Chap. 2, the meanings of the terms stress, strain, elastic limit, yield point, and ultimate strength were reviewed. Students should have these basic definitions clearly in mind before proceeding with this chapter.

The actual stress that a material develops at any one time is caused or induced by the loads acting on the material at that time. Hence, such stress is often called the *induced stress*. Note that the induced stress is not fixed as is the elastic limit or ultimate strength, but varies according to the magnitude of the applied loads. Also the induced stress can be below or above the elastic limit, very small or as much as the ultimate strength of the material.

32. Limit Stress, Design Stress, and Factor of Safety. To prevent failure of a structural member or machine part, the induced stress must of course at *all* times be below the ultimate strength of the material. Of equal importance, to prevent permanent deformation, the induced stress must at all times be below the elastic limit and where repeated stress or reversal of stress is frequent the induced stress must never reach the endurance limit. The highest value of the induced stress which the part can develop without either permanent deformation or failure is called the *limit stress*.

The *design stress*, or *allowable stress*, is that stress value selected for use in designing a structural member or machine part. What value should be chosen for the design stress? A value too low means excessive costs; one too high means danger of failure. Theoretically, the design stress can be as high as the limit stress. Practically, it should be much lower, because of the two factors mentioned in Chap. 1:

1. It is never possible to determine exactly the magnitudes of the applied forces.

2. One sample of a material varies in strength from another.

The induced stress is likely, therefore, at times to be greater than the design stress, and unless the design stress is made sufficiently low, the induced stress may reach or exceed the limit stress.

If the limit-stress value is twice as large as the selected design-stress value, it is said that the designer is using a *factor of safety* of 2. Hence, factor of safety means the number of times that the value of the limit stress exceeds that of the design stress. Mathematically,

$$N = \frac{s_l}{s_d} \tag{8}$$

where N equals the factor of safety, s_l the limit stress of the material, and s_d the design stress.

Formerly, it was common practice to use the ultimate strength in the above equation instead of the limit stress. This practice, however, did not give the real safety factor.

In design, a value of the factor of safety is first decided upon. Then with the limit stress of the material known, the design stress can be calculated, for

$$s_d = \frac{s_l}{N}$$

In machine design the factor of safety chosen for any given material may vary from 1.5 to as high as about 6. Why is this so? The answer is found in the manner of load application. When the loads are steady (static), their magnitudes can be determined with considerable accuracy, and hence the factor of safety can be kept low. On the other hand, when loads are variable and when in addition impact may be present, the value of these forces cannot be determined with such accuracy. In these cases the value of the factor of safety should be high. It follows that a high factor of safety means a low design stress. Furthermore, variable loads mean repeated load applications which may mean that the endurance limit is used as the limit stress. Since this value is lower than the elastic limit, a lower value of the design stress for any given factor of safety would be obtained.

From the foregoing, it can be seen that in machine design a value of the design stress considerably below that used in structural design is the rule. Values of 8000 to 10,000 psi for steel are not uncommon in machine design, whereas, in structural design values of 18,000 to 20,000 psi are often used.

We shall now investigate the types of variable loading encountered in machine design and the stresses that are so produced.

33. Dynamic Loads. Variable loads are known as *dynamic loads* and the stresses produced as *dynamic stresses*. In machine design, dynamic stresses are of far greater importance than static stresses. Dynamic stresses are of two types.

The first type consists of those stresses produced by variable outside loads. Let us call to mind the crane problem of Sec. 25; illustrated in Fig. 35. You will recall the conclusion that at rest or uniform motion up or down, the tension on the cable holding the weight is equal to the weight, but that the tension increases as the weight is accelerated upward and decreases as it is accelerated downward. Knowing the value of the acceleration, we are able to calculate the additional load on the cable because of this acceleration.

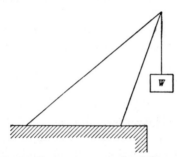
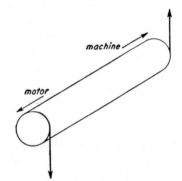

Fig. 35. Crane—an example of dynamic loads.

Fig. 36. Drive shaft—another example of dynamic loads.

Another example of dynamic loading is found in a drive shaft connecting a motor and some machine to be driven (Fig. 36). At the near end of the shaft the motor creates a clockwise moment (torque) and at the far end the friction of the machine and the work being accomplished resists the tendency toward rotation by a counterclockwise torque. The value of the forces producing the torques continually change as the speed or resistance changes.

Large structures, such as buildings, bridges, and the like, are also subjected to variable loads, but proportionately to a lesser extent. In buildings, for example, much of the loading on the beams, girders, and columns is dead load, that is, the weight of the structure itself.

34. Inertia Loads. The second type of dynamic loads are known as *inertia loads*. These loads, peculiar to moving parts only, are caused by the inertia of the part itself, not by outside forces. An example is found in a flywheel such as that discussed in Sec. 30 (Fig. 37). According to Newton's first law (law of inertia), every particle in the wheel tends to fly off in a straight line tangent to the circle of rotation when the wheel is rotating at either uniform or accelerated speed. This centrifugal tendency is greatest at the rim where the linear speed is the greatest and most of the mass is concentrated. The result is a tension in the web and rim of the wheel following radial lines of stress as well as a tension in the rim following circumferential lines of stress. For the determination of the stress value, the centrifugal force formula of physics is used as a starting point.

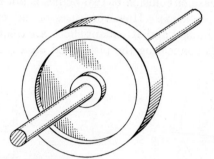

Fig. 37. Flywheel—an example of inertia loads.

35. Impact Loads. When loads are suddenly applied, shock is present to a greater or less degree. An extreme case of impact is that of the collision of two bodies, such as that of a hammer and an anvil. When no motion is imparted by the striking body to the body which is struck, all the kinetic energy of the moving body is absorbed by the stationary body. The hammer strikes the anvil with considerable velocity and must momentarily come to rest before again springing back. The distance traveled by the hammer from the instant it contacts the anvil until it reaches the state of rest is the amount of deformation of the hammer, the anvil, and its base—an exceedingly short distance.

If we assume that the elastic limits of the hammer, anvil, and base are not exceeded, then the negative acceleration of the hammer as it comes to rest is constant, and we can make use of one of the motion formulas for constant accelerated motion for bodies starting from or coming to rest. As given in Sec. 30,

$$v^2 = 2as \quad \text{and} \quad a = \frac{v^2}{2s} \tag{6}$$

where v in this case is the initial velocity and s is the distance traveled by the hammer from contact to rest. Since s is very small, a will be very large. Also Newton's second law can be applied. As explained in Sec. 25,

$$\frac{F}{W} = \frac{a}{g} \quad \text{and} \quad F = \frac{Wa}{g} \tag{4}$$

where in this case F is the force of the hammer on the anvil, W the weight of the hammer, and g, as always, the acceleration of gravity.

But a, as we have just seen, is very large, and hence the force F is very large also, even though the hammer itself may be light in weight.

The extreme case of collision of two bodies is not met with often in machine-design problems. However, cases of suddenly applied loads with a degree of impact and attendant shock are frequent. For example, the wrist pin connecting the piston rod on a reciprocating steam

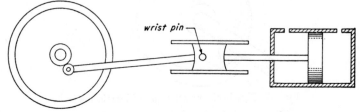

Fig. 38. Impact on wrist pin of steam engine.

engine (Fig. 38) is subjected to impact twice for every revolution of the wheel, that is, at each dead-center position. To reduce the impact, it is important that the joint be made as tight as is proper for such a connection. The drive shaft of an automobile is another example of a machine part which must often withstand considerable shock, especially when an inexperienced driver is at the wheel. To a greater or lesser degree, shock is present whenever the gears are shifted to change the speed of the vehicle or to reverse its direction.

Because of the very large stress values that are induced by impact forces, it can be seen why less tough and more brittle materials will shatter under these loads. The factor of safety assigned to the more brittle materials when they are to be used to resist suddenly applied loads with some degree of shock may be several times the value assigned if the loads were to be applied gradually. Better still, the use of brittle materials where shock is present should be avoided.

36. Stress Concentration. Another factor, hitherto unmentioned, which affects the value of the factor of safety and, hence, the design stress particularly in problems of machine design is the matter of *stress concentration.*

Figure 39a represents a plate with a hole subjected to tensile loads P. Figures 39b and c show the stress at the solid section AA and the discontinuous section *BB*, respectively.   Since the loads are axial, every part of section *A A* is equally stressed, whereas the stress in section *BB* rises sharply at the vicinity of the hole and reaches a maximum at its edges.

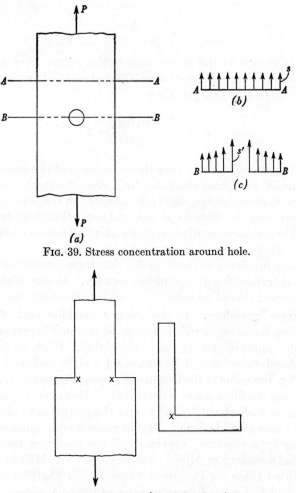

Fig. 39. Stress concentration around hole.

Fig. 40. Stress concentration at reentrant corners.

In Fig. 40 two more examples of stress concentration are illustrated. The crosses indicate the points of increased stress in the plate and the angle.   To relieve the stress at such places, sharp inner corners should be avoided by specifying fillets in the design drawings.

From the illustrations it can be seen that, wherever the *line of stress* is deflected from a regular path, the stress increases at the deflection.

This condition is known as stress concentration. Referring again to Fig. 39c, s represents the stress value if the line of stress were not deflected and s' represents the maximum value owing to the deflection. Then

$$\frac{s'}{s} = K \tag{9}$$

where K is known as the *stress-concentration factor*. Let us assume that the stress in the solid section is 5000 psi and the maximum stress at the hole edge is 10,000 psi; then

$$K = \frac{s'}{s} = \frac{10,000}{5000} = 2$$

Actually, because of the fact that the stress around the hole decreases as this part of the plate elongates to a greater extent (more stress means greater deformation), the true value of the stress-concentration factor is less than the theoretical one, and corrections may be made in design. The stress-concentration factor also varies somewhat with the material.

As in cases involving variable loads, the stress-concentration factor cannot be determined with complete accuracy. Hence, higher values of safety factors should be used.

37. Design Procedure. In designing a machine part, the direct load, bending moment, or twisting moment (torque) is first considered as statically applied and its value calculated. If shock is present, this calculated value should be increased and, if fatigue is present, theoretically, the value of the design stress (when calculated) should be lowered (since the limit stress is decreased). However, in practice the designer takes a short cut by increasing the static-load value to take care of both shock and fatigue. This increase is accomplished through multiplying by a constant, varying with the condition that obtains. The American Society of Mechanical Engineers (ASME) in its publication entitled Code for Design of Transmission Shafting offers constants K_m and K_t for bending moments and twisting moments, respectively, for shafting. The conditions vary as shown in Table 3.

The next steps are to select the material, decide upon a factor of safety, determine the limit stress, and calculate the design stress. Then, if the proposed part is to have holes, grooves, or sharp bends, stress concentration must be taken into account by an appropriate reduction of the design stress. When all these things have been done, the designer is ready to apply the proper design formulas.

TABLE 3

Nature of loading	Values for K_m	Values for K_t
Stationary shafts:		
Gradually applied loads.........................	1.0	1.0
Suddenly applied loads.........................	1.5–2.0	1.5–2.0
Rotating shafts:		
Gradually applied or steady loads.................	1.5	1.0
Suddenly applied loads, minor shocks only..........	1.5–2.0	1.0–1.5
Suddenly applied loads, heavy shocks..............	2.0–3.0	1.5–3.0

This would be good design procedure.   However, we shall simplify this procedure somewhat in the pages that follow.   During the explanations of design formulas, we shall neglect the shock and fatigue factors, but for problem solving, the factor of safety or design stress given will have been adjusted for shock, fatigue, and stress concentration, if necessary.

PROBLEMS

1. Under what conditions would the elastic limit of the material be used as the limit stress value?

2. How does induced stress differ from design stress?   When is the value of each the same?

3. The drive shaft of an automobile is to be made of a certain alloy steel.   The material has an ultimate strength of 150,000 psi, an elastic limit of 115,000 psi, and an endurance limit of 65,000 psi.   Calculate the proper design stress, assuming a factor of safety of 3.

4. What are the reasons for the practice of using a lower design stress in machine design as compared to structural design?

5. What is meant by inertia loads?

6. Explain why very high stresses are likely to occur in bodies when in collision.

7. Why are sharp bends potential points of weakness in machine parts?   What can be done to minimize the effect of such bends?

8. In the design of a certain machine part, allowance is made for a condition of stress concentration in the calculation of the design stress as well by an increase in the factor of safety.   Explain.

9. How are conditions of shock or fatigue commonly taken into account in applications of design formulas?

CHAPTER 7

POWER AND POWER TRANSMISSION

38. Work and Power. Although a man or group of men can excavate for a building with picks and shovels, a power shovel is used nowadays because by its use the hole is dug much faster. When time is not considered, both man and machine may be thought of as being able to do the same amount of work. On the other hand, work per unit of time is *power* and, since the machine does the work in much less time, we say that it is many times more powerful than the man.

As discussed in Chap. 5, mechanical work is the product of force and the distance through which the force acts with both force and distance measured in the same direction.

$$\text{Work} = Fs \tag{5}$$

Power is this work divided by time, or

$$\text{Power} = \frac{Fs}{t}$$

If an engine is able to accomplish 550 ft-lb of work in 1 sec or 33,000 ft-lb in 1 min, it is said to be doing work at the rate of 1 horsepower (hp).[1] Therefore, the power rating in horsepower H is found by dividing the power in foot-pounds per second by 550 or the power in foot-pounds per minute by 33,000:

$$H = \frac{Fs}{550t}$$

where t represents time expressed in seconds;

$$H = \frac{Fs}{33,000t} \tag{10}$$

where t represents time expressed in minutes.

[1] The horsepower unit was established by James Watt, the engineer credited with the invention of the steam engine. Watt was commissioned to rate the steam engines used in pumping water from the mines in England in terms of the equivalent number of horses formerly used to do this job.

39. Power, Force, and Speed.   Steam and gas engines, electric motors, and other types of engines are used to do mechanical work through machines of various kinds.   Each type of machine, such as a pump, hoist, or lathe, has its own specialized function, and, in order to use the power delivered by the engine most effectively, the machine is designed to change the speed and magnitude of the force.   Let us take the example of a crane or hoist.   A hoist of 1 hp capacity can lift a weight of 1000 lb a distance of 33 ft in 1 min, since

$$H = \frac{Fs}{33,000t} = \frac{1000 \times 33}{33,000 \times 1} = 1 \qquad (10)$$

If the hoist were required to lift a weight of 2000 lb this distance of 33 ft, it would take twice the length of time (2 min) because the machine cannot work any faster than at the rate of 1 hp.

$$H = \frac{Fs}{33,000t} = \frac{2000 \times 33}{33,000 \times 2} = 1 \qquad (10)$$

The increase in force and the decrease in speed necessary in the second case are accomplished by the operation of shifting gears (changing the gear ratio) in the machine.

From the above we can state this fundamental principle.   For constant power, as the force increases, the speed decreases proportionately, and vice versa.   Looking at it another way,

$$\text{Power} = \frac{Fs}{t}$$

but s/t = speed or velocity v, and hence

$$\text{Power} = Fv$$

40. Power, Torque, and Speed.   In the previous section we have thought of motion mainly as straight-line motion.   Now let us apply the principle to the motion of rotation.   Figure 41 illustrates a shaft transmitting power by means of belts and pulleys.   The power delivered by the motor at one end of the system goes to the machine at the other end and is used to accomplish mechanical work.   The moment causing the shaft to rotate is the *twisting moment* or *torque*, whereas the moment resisting this rotation is the *resisting moment*.

In Fig. 42 the force P_1 represents the net force of the belt on the pulley circumference caused by the pull from the motor.   In one revolution, P_1 travels once around the circumference of the pulley,

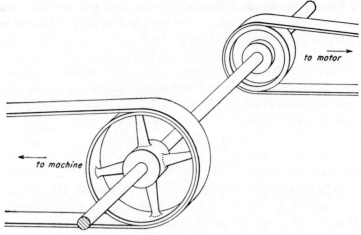

Fig. 41. Power transmission by shaft and pulleys.

or a distance of $2\pi r$, where r is the radius. Hence,

$$\text{Work done in 1 revolution} = P_1 \times 2\pi r$$

If n represents the number of revolutions per minute (rpm), then

$$\text{Work done in 1 min} = P_1 \times 2\pi r \times n$$

But as previously explained, work done per unit of time is power. Also $P_1 \times r$ is the torque T. Then, by substitution, we have

$$\text{Power} = T \times 2\pi \times n$$

To use this equation for finding power in horsepower, we must first change the work from inchpounds (since r is in inches) to foot-pounds by dividing by 12, and then divide by 33,000. Therefore,

Fig. 42.

$$H = \frac{T \times 2\pi \times n}{33,000 \times 12}$$

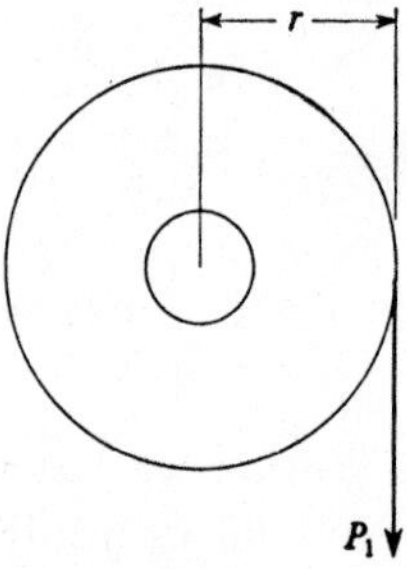

By substitution of a numerical value for π, the equation can be reduced to

$$H = \frac{Tn}{63,000} \tag{11}$$

This formula is used to solve for H, T, or n when the other two terms are known.

In the previous section it was brought out that, for constant power, as the force increases the speed must decrease, or vice versa. From Eq. (11) it can be seen that a similar relationship exists between power, torque, and speed of rotation, or

$$\text{Power} = Tn$$

Again for constant power, as the torque increases, the rotational speed decreases proportionately, and vice versa. In fact, there can be an infinite number of variations of torque and rotational speed for any given value of the power as long as the product of T and n is equal to that value of the power.

Immediately after starting, an automobile is rapidly picking up speed (large acceleration). During this period a large force (and hence a large torque) is needed, because, again according to Newton's second law, force is proportional to the acceleration produced. Also during the period of acceleration the rotational speed is small. On the other hand, when the vehicle has attained a relatively high speed and is traveling at constant speed, force is needed to overcome frictional resistance only. During this latter period torque is small and rotational speed is large.

The varying drive-shaft speeds in an automobile are accomplished by the gearshift mechanism or by a fluid clutch as well as by the accelerator (gas pedal) to change the speed of the motor. Even though the engines of trucks in general can develop greater horsepower as compared with passenger automobiles, most trucks move more slowly because of the great force needed to propel their greater weights.

The relationships of power, force (or torque), and speed just discussed are fundamental to the design of all machine parts where motion is present. It is most necessary for the beginner in machine design to acquire a thorough understanding of these principles.

Illustrative Example. The drive shaft of an automobile, 2 in. in diameter, is driven by a 120-hp motor.

a. Calculate the torque on the shaft when the engine is developing its full horsepower and the shaft is rotating at a speed of 2000 rpm.

b. A 9-in.-diam gear is keyed to the shaft. What force is developed at its circumference?

c. What horsepower is delivered by the driving wheels if the efficiency of the driving mechanism is 90 per cent?

a.

$$T = \frac{63,000H}{n} = \frac{63,000 \times 120}{2000} = \frac{7,560,000}{2000} = 3780 \text{ lb-in.} \quad (11)$$

b.

$$T = P_1 r$$

$$P_1 = \frac{T}{r} = \frac{3780}{4.5} = 840 \text{ lb}$$

c.

$$\text{Efficiency} = \frac{\text{work output}}{\text{work input}}$$

Since horsepower is a measure of the quantity of work done per unit of time, and since the time is identical for both horsepower output and horsepower input, horsepower can be used for work in the efficiency equation.

$$0.90 = \frac{\text{hp output}}{120}$$

$$\text{hp output} = 0.90 \times 120 = 108$$

41. Drawing Power from the Shaft. Students are often confused as to the distribution of the power among various gears and pulleys on a shaft. In Fig. 43*a* the shaft is directly connected to the source

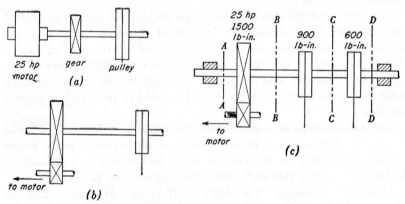

Fig. 43. Examples of power distribution.

of power as an electric motor, turbine, or internal-combustion engine. The gear and pulley, conventionally represented as shown, are transmitting the 25 hp supplied to, let us say, two machines. Although the amount of power transmitted to each machine may differ, the *sum of the two amounts must equal 25 hp.* If this were not true, the shaft would continually accelerate either positively until something broke or negatively until it came to a stop. Only during periods of speeding up and slowing down are the horsepower delivered to the shaft and the horsepower delivered by the shaft unequal. Case *b* shows the gear on the shaft being supplied with 25 hp from a smaller gear. This time the pulley draws all the 25 hp to run a machine. Case c differs from case b in that two pulleys are now drawing off the power, dividing it between them. As in case a, the sum of the two amounts must equal the 25 hp supplied through the gears.

42. Varying Torque Values in a Shaft. If in case c the values of the horsepower delivered by the motor through the gears and that drawn off by each pulley are known, the torques can be calculated for any given rotational speed of shaft according to

$$T = \frac{63,000H}{n} \tag{11}$$

These torques are assumed to be 1500, 900, and 600 lb-in., as shown in the figure. It is to be noted that the directions of the twisting moments of the two pulleys are the same but opposite to the direction of the twisting moment of the gear.

At rest or constant speed, the *net* tendency to the left of any cross section of the shaft to rotate the shaft in one direction (clockwise or counterclockwise) is balanced by the *net* tendency to the right of that section to rotate in the opposite direction. Furthermore, this section must be sufficiently strong to resist these equal and opposite moments.

For example, at section AA there is no twisting moment to the left and the twisting moments to the right (when added algebraically) also equal zero.

$$1500 - 900 - 600 = 0$$

At section BB the twisting moment to the left is 1500 lb-in. to balance the sum of the two moments to the right.

$$1500 = 900 + 600$$

Hence, the torque at section BB is 1500 lb-in. At section CC the net twisting moment to the left must again be balanced by that to the right.

$$1500 - 900 = 600$$

The torque at section CC is, therefore, 600 lb-in. At section DD, similar to section AA, the twisting moments to the left when added algebraically equal zero to balance the absence of a moment to the right.

$$1500 - 900 - 600 = 0$$

Therefore, it can be stated that, at any section of the shaft, the torque that must be resisted by that section is equal to the algebraic sum of the twisting moments to the left or right of that section. In design the greatest torque value must be used because the induced stress is proportional to the torque, as we shall see in Chap. 8.

PROBLEMS

1. The bottom of a cylindrical water tank, 10 ft in diameter and 12 ft high, is located 50 ft from the ground level. A 1-hp electric motor is used to run a pump to deliver water to the top of the tank from a well where the water surface is 15 ft below ground. The efficiency of the pump is 75 per cent. Calculate the time required to fill the tank, assuming that water weighs 62.4 lb per cu ft.

2. A 24-in.-diam circular saw makes 800 rpm. Its tangential thrust (force at the circumference) is 25 lb. Calculate:

  a. The torque transmitted by the shaft

  b. The horsepower delivered

  c. The tooth velocity in feet per minute

3. A 1-in.-diam shaft can resist safely a torque of 1600 lb-in. Calculate the horsepower it can deliver when revolving at a speed of 60 rpm.

4. A 30-in.-diam rotor on a shaft is acted on by a tangential force of 160 lb at the rotor circumference. At what speed must the shaft rotate to deliver 5 hp?

5. In a Prony brake test of a motor as shown in the figure, the spring balance at the end of the brake arm reads 12 lb. A tachometer reading shows that the motor is rotating at 1140 rpm. Calculate the horsepower delivered by the motor. *Hint:* This condition is similar to one in which the brake arm rotates with the pulley and exerts a moving load of 12 lb at its outer end.

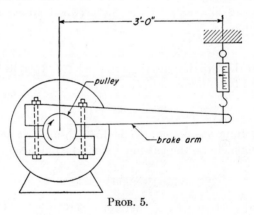

PROB. 5.

6. A testing machine with a capacity of 100,000 lb is to be run by an electric motor. The movable head, which accommodates one set of grips for the specimens, is designed for a maximum speed of 6 in. per min (see figure on page 65). Friction accounts for a loss of 50 per cent of the power.

  a. What horsepower motor is necessary?

  b. What is the value of the torque developed in the motor shaft when running at full horsepower and at a speed of 1720 rpm?

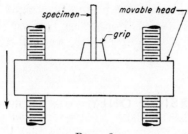

PROB. 6.

7. A 40-hp marine engine drives a 25-in. propeller at a speed of 700 rpm. What would be the speed if a 30-in. propeller were substituted? *Hint:* The torque varies as the area of the propeller circle.

8. A loaded truck weighing 10 tons is traveling up a hill at a speed of 20 miles per hour (mph). The grade of the incline is 10 per cent (consider a 10-ft rise for every 100 ft on the slope). Frictional losses, including wind resistance, account for a power loss of 70 per cent. What horsepower is delivered by the engine?

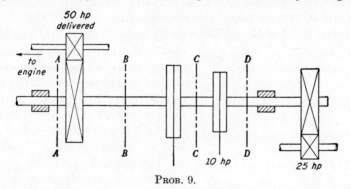

PROB. 9.

9. A 50-hp engine is delivering power to the shaft through the pinion and spur gear as shown in the figure.

a. How much horsepower is being drawn off by the left-hand pulley?

b. What are the torque values at sections AA, BB, CC, and DD if the shaft is rotating at 300 rpm?

CHAPTER 8

SHAFTS IN TORSION ONLY

43. The Stress of Torsion. A shaft develops a stress of torsion when resisting a twisting movement. For metallic materials torsional failure takes the form of a smooth break in a plane at right angles to the axis of the shaft. This indicates a shear failure. One part of the shaft has slid past the other part with a rotational motion about the center of gravity point of the cross section. Hence, this point is the neutral point of the section, and its locus is the neutral axis of the shaft. See Fig. 44 and note that the neutral axis coincides with the longitudinal axis of the shaft. The more ductile steels show greater torsional deformation prior to breaking than those steels with greater carbon content or which have been hardened by heat-treatment. Some metals, for example, aluminum or copper, are so soft and ductile that they will not shear even after the specimen has made ten or more complete revolutions.

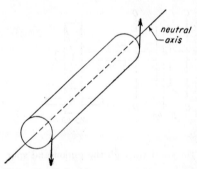

FIG. 44. Coincidence of shaft axis and neutral axis.

Twisting a wood shaft to failure results in a series of longitudinal breaks caused by sliding in planes *parallel* to the axis. If the ends of the wood shaft are smooth before the application of the twisting loads, the ends become jagged after the appearance of the longitudinal breaks. This verifies the fact that sliding has taken place. Wood shafts fail differently from metal shafts because wood is weakest in the direction parallel to the grain.

Torsion, therefore, is shear acting in both perpendicular and parallel directions to the axis of the member being twisted.

44. Shafting—Materials and Sizes. Shafting is usually of mild or medium plain carbon steel. The carbon content ranges from about

0.25 per cent to about 0.40 per cent (AISI 1025 to 1040). Alloy steels are sometimes used where extra strength or hardness is called for, with nickel or chromium as the common alloying element.

Shafts are made either by the hot-rolled or the cold-rolled process. Hot-rolled shafts are machined to size (turned) and then finished. Because of the work-hardening effect of cold rolling, such shafts are stronger and less ductile than the hot-rolled product. However, cold-rolled shafts have residual stresses near the circumference as a result of the rolling.

Stock sizes for shafts, as given in manufacturers' catalogues, are as follows:

Diameters, In.	Diameter Increments, In.
Up to 2	$\frac{1}{16}$
2 to 5	$\frac{1}{4}$
5 to 8	$\frac{1}{2}$

45. Design of Solid Shafts in Torsion Only. When the horsepower to be delivered and the rpm are known, the torque that the shaft must resist is calculated from Eq. (11). Then the *size of shaft* can be determined by the use of the *torsion formula*[1]:

$$T = s\frac{J}{c}$$

where T = maximum torque (twisting moment), lb-in.

sJ/c = resisting moment developed by the shaft

s = appropriate design stress of the material

J = polar moment of inertia of the section (cross section), that is, moment of inertia about the point where the neutral axis intersects the section

c = distance from the neutral axis to the extreme fibers; in circular shafts this distance is the radius

The formula for the polar moment of inertia of a circular section about its center is $d^4/32$ and the distance c equals $d/2$. Making these substitutions in the torsion formula, we obtain

$$\frac{T}{s} = \frac{\pi d^4}{32} \times \frac{2}{d}$$

which simplifies to

$$\frac{T}{s} = \frac{\pi d^3}{16} \quad \text{or} \quad \frac{T}{s} = \frac{d^3}{5.10} \tag{12}$$

[1] For a derivation of this formula, the student is advised to consult elementary texts on strength of materials.

This equation can be used to determine the torque or the stress developed in resisting any given torque for any known size of shaft, as well as for the design of a shaft when the other terms are known. In problems involving the selection of a shaft size, it will be found convenient to solve Eq. (12) for d before making numerical substitutions,

$$d = \sqrt[3]{\frac{16T}{\pi s}} = \sqrt[3]{\frac{5.10T}{s}}$$

which can be further reduced to

$$d = 1.72 \sqrt[3]{\frac{T}{s}} \tag{13}$$

When the horsepower to be delivered and the rpm are given, the size of shaft can be solved directly by use of Eq. (14), as explained below, thus eliminating the intermediate step of solving for the torque.

$$T = \frac{63,000H}{n} \tag{11}$$

and

$$d = 1.72 \sqrt[3]{\frac{T}{s}} \tag{13}$$

By making substitutions for T in (13) from (11), we get

$$d = 1.72 \sqrt[3]{\frac{63,000H}{ns}}$$

and, simplifying,

$$d = 68.5 \sqrt[3]{\frac{H}{ns}} \tag{14}$$

Because of the fact that most shafts are subjected to bending stress in addition to torsional stress, the application of this formula is limited. Also since this formula involves the cube root of a very small number, it is somewhat difficult to handle.

Illustrative Example. Select the most suitable round steel section to be used as a shaft to transmit 100 hp at a speed of 250 rpm from the transmission of an automobile truck to the rear axle. Assume a design stress of 10,000 psi.

$$T = \frac{63,000H}{n} = \frac{63,000 \times 100}{250} = 25,200 \text{ lb-in.} \tag{11}$$

$$d = 1.72 \sqrt[3]{\frac{T}{s}} = 1.72 \sqrt[3]{\frac{25,200}{10,000}} = 1.72 \times 1.36 = 2.34 \text{ in.} \tag{13}$$

or, using (14),

$$d = 68.5 \sqrt[3]{\frac{H}{ns}} = 68.5 \sqrt[3]{\frac{100}{250 \times 10,000}} = 68.5 \sqrt[3]{0.0000400} = 2.34 \text{ in.}$$

Use 2.50 in. as the next larger standard size.

46. Design of Hollow Shafts in Torsion Only.   How can a shaft be made lighter without increasing the stress; or how can a shaft be strengthened without adding to its weight and hence to its cost? These desirable ends are achieved by means of hollow shafts.

From strength of materials we remember that the stress in a shaft section is zero at the neutral axis (center of the circle), increases proportionately to the distance from the center, and reaches a maximum at the circumference.   Bearing this fact in mind, we realize that the material near the center does much less resisting of the torque than

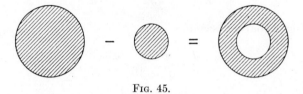

Fig. 45.

the material near the circumference and, hence, is largely wasted.   In a hollow shaft, the material that does little resisting is eliminated and material is added where it does more good.   A hollow shaft, therefore, is stronger than a solid shaft of equal cross-sectional area.   Where large-diameter shafts are needed, a considerable saving is effected by the use of a hollow section.

Both for area and polar moment of inertia calculations, such a section may be considered as the difference between a large solid section and a smaller solid section, as shown in Fig. 45.   Remembering that $J = d^4/32$ for a solid circle, then for a hollow circle

$$J = \frac{\pi d^4}{32} - \frac{\pi d_1^4}{32}$$

where d and d_1 are the outside and inside diameters, respectively.   This expression can be factored to obtain

$$J = \frac{\pi(d^4 - d_1^4)}{32}$$

Let the symbol q be used for the ratio of the inside diameter to the outside diameter,

$$\frac{d_1}{d} = q$$

or

$$d_1 = dq$$

Now let us substitute dq for its equal d_1 in the polar moment of inertia expression.

$$J = \frac{\pi(d^4 - d^4 q^4)}{32}$$

which can be factored to

$$J = \frac{\pi d^4(1 - q^4)}{32}$$

Since $c = d/2$,

$$\frac{J}{c} = \frac{\pi d^4(1 - q^4)}{32} \times \frac{2}{d}$$

which by cancellation becomes

$$\frac{J}{c} = \frac{\pi d^3(1 - q^4)}{16}$$

and the torsion formula results in

$$\frac{T}{s} = \frac{\pi d^3(1 - q^4)}{16}$$

which upon further simplification becomes

$$\frac{T}{s} = \frac{d^3(1 - q^4)}{5.10} \tag{15}$$

Again as in the case of solid shafts, the above expression can be solved for d.

$$d = 1.72 \sqrt[3]{\frac{T}{s(1 - q^4)}} \tag{16}$$

Remember that Eqs. (15) and (16) are to be used for hollow shafts only.

If the hole in a shaft is made larger and larger without adding more material at the circumference, the wall will become very thin. Such a section is in danger of collapse when in use. For this reason and for the sake of simplicity, the ratio of inside diameter to outside diameter of hollow shafts is rarely made other than 1:2 and 1:3, that is,

q equals 0.500 or 0.333. The following table will be of value when using Eqs. (15) and (16).

q	q^4	$1 - q^4$
0.5	0.0625	0.938
0.333	0.0123	0.988

Illustrative Example. Investigate a 2-in.-ID and 4-in.-OD steel shaft to determine the horsepower it can transmit safely to the propeller of a ship. The speed of the shaft is 100 rpm. The limit stress of the steel in shear is 24,000 psi, and a factor of safety of 3 is specified.

$$s_d = \frac{s_l}{N} = \frac{24,000}{3} = 8000 \text{ psi} \tag{8}$$

$$T = \frac{sd^3(1 - q^4)}{5.10} = \frac{8000 \times 4 \times 4 \times 4 \times 0.938}{5.10} \tag{15}$$
$$= 94,200 \text{ lb-in.}$$

$$H = \frac{Tn}{63,000} = \frac{94,200 \times 100}{63,000} = 150 \tag{11}$$

PROBLEMS

1. State one advantage and one disadvantage of hot-rolled as compared to cold-rolled steel for shafting.

2. What is the value of the torque that a 1$\frac{5}{16}$-in.-diam shaft resists when stressed to its design stress of 10,000 psi?

3. A 1$\frac{1}{2}$-in.-diam shaft resists a force of 850 lb at the circumference of a 16-in.-diam pulley. Calculate the torsional stress induced in the shaft.

4. Determine by calculation the required diameter of a shaft to resist a torque of 12,000 lb-in. The design stress is 10,000 psi.

5. A cold-rolled steel shaft is required to transmit a torque of 40,600 lb-in. The limit stress is 22,500 psi and a factor of safety of 2.5 is specified.

a. Calculate the required diameter.

b. What commercial size should be used?

6. What is the horsepower that a 2$\frac{3}{4}$-in.-diam line shaft can develop safely when rotating at 150 rpm? The design stress of the material is 8000 psi.

7. A 100-hp motor is to activate a centrifugal pump through a direct connection. The shaft is 2 in. in diameter. Calculate the speed of the pump when the shaft is stressed to 10,000 psi.

8. A shaft is to rotate at 100 rpm and transmit a torque of 33,800 lb-in. A design stress of 10,000 psi is specified. What commercial size of shaft should be chosen?

9. Calculate the required diameter of the drive shaft for an automobile when transmitting 80 hp at 2500 rpm. Use a design stress of 8000 psi. What commercial size should be used?

10. A four-cycle eight-cylinder gasoline engine is designed to develop 115 hp. The crankshaft rotates at 1650 rpm. Calculate:

    a. The torque that the crankshaft must resist.

    b. The most suitable commercial size to use. The design stress is 10,000 psi.

11. A hollow circular shaft is 4½ in. OD and 1½ in. ID. Calculate the maximum horsepower it can transmit safely when rotating at a speed of 170 rpm. The design stress is 10,000 psi.

12. The shaft of a boat must transmit 200 hp to the propeller at a speed of 350 rpm. A hollow shaft is to be used with a ratio of outside to inside diameter of 2:1. Calculate the most suitable size section, using a design stress of 9000 psi.

13. A turbine in a hydroelectric plant is able to generate 1200 hp when the blades turn at a speed of 160 rpm. Calculate the most suitable commercial size hollow shaft for this purpose with inside diameter one-third of the outside diameter. The limit stress is 26,000 psi and a factor of safety of 3 is specified.

CHAPTER 9

SHAFTS IN BENDING AND TORSION

47. Loads of Gears and Pulleys. In addition to the induced stress of torsion described in Chap. 8, all shafts act as beams and, in consequence, must resist bending stresses. Sometimes the bending stress is relatively small and can be neglected in design; more often it must be taken into account.

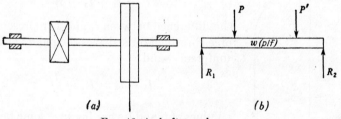

(a)   *(b)*

FIG. 46. A shaft as a beam.

The weight of the shaft itself is a uniformly distributed load throughout the length, whereas the various gears and pulleys give rise to concentrated loads. Figure 46a represents a shaft beam with the supports (bearings) located at the ends. Figure 46b represents the loading diagram with gear and pulley concentrated loads as P and P', respectively, uniform shaft load as w (pounds per linear foot), and the supports as reactions R_1 and R_2. The concentrated loads are composed of, first of all, the static loads (dead weights) of gear or pulley and, second and of greater importance, factors called the *tooth thrust* in the case of the gear and *belt pull* in the case of the pulley.

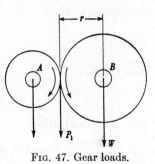

FIG. 47. Gear loads.

Tooth thrust is the result of the torque transmitted from a gear to a mating gear. In Fig. 47 let us assume that shaft A is driven

73

by shaft B and that the torque of B is T, made up of the force on the gear teeth P_1 multiplied by the radius of the gear r.

$$T = P_1r$$

and

$$P_1 = \frac{T}{r}$$

The greatest concentrated load P occurs when the tooth thrust P_1 is considered as acting in the same direction as the static load W, that is, vertically downward.

$$P = P_1 + W \qquad \text{and} \qquad P = \frac{T}{r} + W \qquad (17)$$

In the case of the pulley (Fig. 48), t_1 represents the tension of the tight side of the belt and t_2 that of the slack side. Again the greatest concentrated load occurs when all forces are in the same direction and

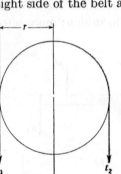

$$P = t_1 + t_2 + W \qquad (18)$$

The difference between t_1 and t_2 represents the force P_1, which force drives the pulley at its circumference, and

$$t_1 - t_2 = P_1 = \frac{T}{r}$$

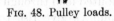

Fig. 48. Pulley loads.

Values of t_1 and t_2 vary because some belts are given more tension than others when installed and all belts stretch through use. However, it is customary to assume that t_1 bears a definite ratio to t_2 as two or three times greater. If we assume that t_1 is twice as large as t_2, then

$$2t_2 - t_2 = P_1$$

and

$$t_2 = P_1$$

also since

$$P = t_1 + t_2 + W \qquad (18)$$
$$P = 2t_2 + t_2 + W = 3t_2 + W$$

and, by substitution,

$$P = 3P_1 + W \qquad \text{or} \qquad P = \frac{3T}{r} + W \qquad (19)$$

Note: For examples in this chapter dealing with pulleys we shall assume a ratio of t_1 to t_2 of $2:1$ unless otherwise specified.

Illustrative Example. Calculate the concentrated loads on a shaft caused by a gear and a pulley which transmit 5 and 8 hp, respectively. The speed of the shaft is 100 rpm. The gear weighs 30 lb and has a pitch diameter of 12 in., whereas the pulley weighs 50 lb and has a diameter of 18 in.

$$T = \frac{63,000H}{n} = \frac{63,000 \times 5}{100} = 3150 \text{ lb-in.} \qquad \textbf{(11)}$$

$$P_1 = \frac{T}{r} = \frac{3150}{6} = 525 \text{ lb}$$

$$P = P_1 + W = 525 + 30 = 555 \text{ lb} \qquad \textbf{(17)}$$

where P is the load caused by the gear.

$$T = \frac{63,000H}{n} = \frac{63,000 \times 8}{100} = 5040 \text{ lb-in.} \qquad \textbf{(11)}$$

$$P_1 = \frac{T}{r} = \frac{5040}{9} = 560 \text{ lb}$$

$$P = 3P_1 + W = 3 \times 560 + 50 = 1730 \text{ lb} \qquad \textbf{(19)}$$

where P is the load caused by the pulley.

48. Maximum Bending Moments. Let us now review that portion of strength of materials dealing with the calculation of the maximum bending moment. We shall consider three cases of simply supported beams.

Case 1. *Concentrated Load at Any Point in the Span.* Since the beam (Fig. 49) is in equilibrium, $\Sigma V = 0$ and $\Sigma M = 0$ apply. Assume a center of moments (point of rotation) about one reaction, say R_2, and solve for R_1 by $\Sigma M = 0$. Clockwise moments are to be given a plus sign and counterclockwise moments a minus sign.

$$R_1 \times l - P(l - x) = 0$$

$$R_1 = \frac{P(l - x)}{l}$$

and taking moments about R_1,

$$P \times x - R_2 \times l = 0$$

$$R_2 = \frac{Px}{l}$$

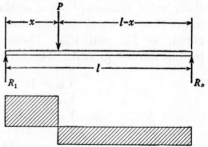

FIG. 49. Case 1: loading and shear diagrams.

It is advisable not to try to remember these general expressions as formulas, but to work out each specific case from $\Sigma M = 0$. It is also advisable to calculate both

R_1 and R_2 by $\Sigma M = 0$ and check the values by adding all the loads algebraically ($\Sigma V = 0$) to see if the result is zero. In this case

$$\frac{P(l - x)}{l} + \frac{Px}{l} - P$$

should equal zero. By simplification,

$$\frac{Pl - Px + Px}{l} - P = P - P = 0$$

Note that the larger of the two reactions is the one nearer the load.

Let us now assume a concentrated load value P of 400 lb acting at a distance x of 6 in. from R_1 on a span l of 17 in. Then, by taking moments about R_2,

$$17R_1 - 400 \times 11 = 0$$
$$R_1 = \frac{400 \times 11}{17} = 259 \text{ lb}$$

and taking moments about R_1,

$$400 \times 6 - 17R_2 = 0$$
$$R_2 = \frac{400 \times 6}{17} = 141 \text{ lb}$$

and, by $\Sigma V = 0$,

$$259 + 141 - 400 = 0$$

The value of the vertical shear at any section of a beam is found by calculating the algebraic sum of the forces to the left (or right) of that particular section. At points of concentrated load, where the shear value changes, both a section an infinitesimal distance to the left of that load and an infinitesimal distance to the right are investigated, as V_{6L} and V_{6R}. These shears, respectively, exclude and include the 400-lb load. The vertical shears in this case are

$$V_{0L} = 0$$
$$V_{0R} = 259$$
$$V_{6L} = 259$$
$$V_{6R} = 259 - 400 = -141$$
$$V_{17L} = -141$$
$$V_{17R} = -141 + 141 = 0$$

These values can be plotted to form a shear diagram, as shown below the loading diagram in Fig. 49.

It will be remembered that the reason for calculating the shears is to determine the point of zero shear (or where the shear changes from plus to minus), at which point the bending moment is a maximum. Maximum bending moment means maximum stress and, therefore, the maximum bending moment is subsequently used in calculating the proper beam size and proportions. If the beam should fail, it would do so at the section of maximum bending moment—hence, the name *dangerous section*. Shear diagrams are not absolutely necessary in the operation of determining the point of zero shear, but are helpful as a means of visualization.

The next step is the calculation of M_{max}, that is, M_6. We shall recall that the value of the maximum bending moment at any section

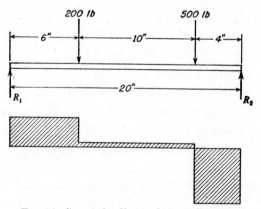

FIG. 50. Case 2: loading and shear diagrams.

of a beam is the algebraic sum of the *moments* to the left (or right) of that section when that section is considered as the center of moments. In this case

$$M_6 = 259 \times 6 = 1554 \text{ lb-in.}$$

Case 2. Two Concentrated Loads at Any Points of the Span. Refer to Fig. 50. Taking moments about R_2 to find R_1,

$$20R_1 - 200 \times 14 - 500 \times 4 = 0$$
$$20R_1 - 2800 - 2000 = 0$$
$$R_1 = 4800/20 = 240 \text{ lb}$$

Taking moments about R_1 to find R_2,

$$200 \times 6 + 500 \times 16 - 20R_2 = 0$$
$$1200 + 8000 - 20R_2 = 0$$
$$R_2 = 9200/20 = 460 \text{ lb}$$

Applying the $\Sigma V = 0$ check,

$$240 - 200 - 500 + 460 = 0$$

Next, values of the vertical shears are calculated:

$$V_{0L} = 0$$
$$V_{0R} = 240$$
$$V_{6L} = 240$$
$$V_{6R} = 240 - 200 = 40$$
$$V_{16L} = 40$$
$$V_{16R} = 40 - 500 = -460$$
$$V_{20L} = -460$$
$$V_{20R} = -460 + 460 = 0$$

Again a shear diagram may be drawn as shown in Fig. 50. We note that the shear is zero (changes sign) at the 500-lb load, which is 16 ft from the left end. Therefore, $M_{max} = M_{16}$.

$$M_{16} = 240 \times 16 - 200 \times 10$$
$$M_{16} = 3840 - 2000 = 1840 \text{ lb-in.}$$

Case 3. *Concentrated Loads and Distributed Loads.* As in cases 1 and 2 we may think of this example (Fig. 51) as a shaft delivering

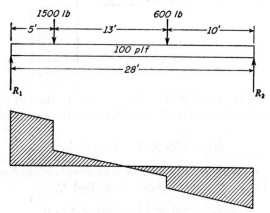

FIG. 51. Case 3: loading and shear diagrams.

power to machines through pulleys or gears at the points of concentrated loads. The uniformly distributed load is the weight of the shaft itself.

For purposes of finding the reactions and bending moments, the entire distributed load is considered as being concentrated at its center of gravity.

$$28R_1 - 1500 \times 23 - 600 \times 10 - 100 \times 28 \times 14 = 0$$
$$28R_1 - 34,500 - 6000 - 39,200 = 0$$
$$R_1 = \frac{79,700}{28} = 2850 \text{ lb}$$
$$1500 \times 5 + 600 \times 18 + 100 \times 28 \times 14 - 28R_2 = 0$$
$$7500 + 10,800 + 39,200 - 28R_2 = 0$$
$$R_2 = \frac{57,500}{28} = 2050 \text{ lb}$$
$$2850 - 1500 - 600 - 2800 + 2050 = 0$$
$$4900 - 4900 = 0$$
$$V_{0L} = 0$$
$$V_{0R} = 2850$$
$$V_{5L} = 2850 - 500 = 2350$$
$$V_{5R} = 2350 - 1500 = 850$$
$$V_{18L} = 850 - 1300 = -450$$
$$V_{18R} = -450 - 600 = -1050$$
$$V_{28L} = -1050 - 1000 = -2050$$
$$V_{28R} = -2050 + 2050 = 0$$

Again it will be recalled from strength of materials that when the point of zero shear occurs between two concentrated loads, its location can be found by dividing the last positive shear by the uniform load per foot, as

$$850/100 = 8.5$$

The point of zero shear is, therefore, $8.5 + 5 = 13.5$ ft from R_1, and

$$M_{13.5} = 2850 \times 13.5 - 1500 \times 8.5 - 100 \times 13.5 \times 6.8$$
$$M_{13.5} = 38,500 - 12,800 - 9180 = 16,500 \text{ lb-ft}$$

49. Fatigue in Shafts. In Sec. 11 fatigue as a cause of failure was discussed and an illustration given of a beam on which the direction

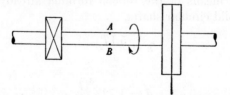

Fig. 52. Rotating shaft causes stress reversals.

of the loads was repeatedly reversing. A shaft beam acts similarly. In the position shown in Fig. 52, the bending stress causes point A to be in compression and point B to be in tension. However, when the shaft has rotated 180°, point B, now on top, is in compression, and point A, now on bottom, is in tension. Thus the stresses of tension and compression caused by bending are continually interchanging for all points on the shaft as it rotates. In addition there are constant changes in the torsional stress as a result of increases and decreases in torque corresponding to changes in speed and load. Because of these factors it is important that the possibility of fatigue failure be taken into account in shaft investigation and design.

50. Bending and Torsional Stresses Combined—The Equivalent Torque. When the induced stresses in a body are all in the same or opposite direction, the resultant, or net, stress can be obtained by an algebraic summation. However, if this condition does not exist, as in shafts in which both torsional and bending stresses are induced, other means of determining the resultant stress must be used. For such shafts varying theories have led to varying methods.

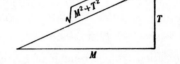

Fig. 53. Graphical determination of equivalent twisting moment.

The method currently recommended by the ASME in its Code for Design of Transmission Shafting is based on what is known as the *maximum shear theory*. In this method a factor which can be called the equivalent twisting moment, or equivalent torque, T_e is calculated by the formula

$$T_e = \sqrt{M^2 + T^2} \tag{20}$$

where M and T are the maximum bending and twisting moments, respectively. The calculated value of T_e is then treated as a pure torque to investigate a shaft of known diameter or to design a new shaft, both by means of the torsion formula already discussed, for example, for solid circular shafts,

$$s = \frac{5.1T}{d^3} \tag{12}$$

or

$$d = 1.72 \sqrt[3]{\frac{T}{s}} \tag{13}$$

The value of T_e can also be obtained graphically. If the two perpendicular sides of a right triangle are drawn to scale to represent M and T, then from geometry we know that the hypotenuse is $\sqrt{M^2 + T^2}$ or T_e (see Fig. 53).

51. The Ideal Torque and Equivalent Bending Moment. A second method of combining bending and torsional stresses is based on what is known as the *maximum principal stress theory*. This is

$$T_i = M + \sqrt{M^2 + T^2} \tag{21}$$

The term T_i is called the ideal torque to distinguish it from the equivalent torque T_e. As in the case of T_e, T_i when calculated is treated as a pure torque and solved by the torsion formula.

A third method (also based on the maximum principal stress theory) is

$$M_e = \tfrac{1}{2}(M + \sqrt{M^2 + T^2}) \tag{22}$$

This combination results in the calculation of an equivalent bending moment M_e. Unlike the two previous methods, M_e is then treated

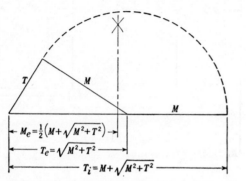

FIG. 54. Graphical determinations of combinations of bending and twisting moments.

as a pure bending moment and the shaft is either investigated or designed as a beam. From strength of materials we remember the flexure formula, namely,

$$\frac{M}{s} = \frac{I}{c} \tag{23}$$

where M = maximum bending moment
s = design stress
I = moment of inertia about the neutral axis of the section (gravity axis perpendicular to the loads)
c = distance from the neutral axis to the extreme fibers (the radius in the case of circular shafts)
This formula used for all beams is now applied to the shaft with M_e substituted for M.

Both the ideal torque and the equivalent bending moment of the second and third methods can also be obtained graphically from the right triangle (see Fig. 54). The graphical method may be used as a rough check of calculations.

Illustrative Example 1. Figure 55 represents a steel shaft supporting a spur gear and a pulley as shown. The gear delivers 50 hp to the

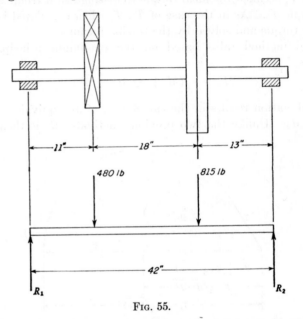

FIG. 55.

pulley at 600 rpm. The gear weighs 40 lb and its pitch diameter is 24 in. The pulley weighs 100 lb and its diameter is 44 in. Design the shaft as a solid section by the method of the equivalent torque. Assume a value of 8000 psi as the design stress and that $t_1 = 2t_2$. Neglect the weight of the shaft.

$$T = \frac{63,000H}{n} = \frac{63,000 \times 50}{600} = 5250 \text{ lb-in.} \quad (11)$$

$$\text{Tooth thrust } (P_1) = \frac{T}{r} = \frac{5250}{12} = 437 \text{ lb}$$

$$P = P_1 + W = 437 + 40 = 477 \text{ lb} \quad (17)$$

Assume P as 480 lb equal to concentrated load of gear.

$$P_1 = \frac{T}{r} = \frac{5250}{22} = 238 \text{ lb}$$

$$P = 3P_1 + W = 3 \times 238 + 100 = 814 \text{ lb} \quad (19)$$

Assume P as 815 lb equal to concentrated load of pulley.

$$42R_1 - 480 \times 31 - 815 \times 13 = 0$$
$$42R_1 - 14,900 - 10,600 = 0$$
$$R_1 = \frac{25,500}{42} = 607 \text{ lb}$$
$$480 \times 11 + 815 \times 29 - 42R_2 = 0$$
$$5280 + 23,600 - 42R_2 = 0$$
$$R_2 = \frac{28,880}{42} = 688 \text{ lb}$$
$$607 - 480 - 815 + 688 = 0$$
$$V_{0L} = 0$$
$$V_{0R} = 607$$
$$V_{11L} = 607$$
$$V_{11R} = 607 - 480 = 127$$
$$V_{29L} = 127$$
$$V_{29R} = 127 - 815 = -688$$
$$V_{42L} = -688$$
$$V_{42R} = -688 + 688 = 0$$

Vertical shear changes sign 29 ft from left end.

$$M_{29} = 607 \times 29 - 480 \times 18 = 17,600 - 8650 = 8950 \text{ lb-in.}$$

$$
\begin{aligned}
T_e &= \sqrt{M^2 + T^2} = \sqrt{(8950)^2 + (5250)^2} \\
&= \sqrt{80,100,000 + 27,600,000} \\
&= \sqrt{107,700,000} = 10,400 \text{ lb-in.}
\end{aligned}
\tag{20}
$$

$$
\begin{aligned}
d &= 1.72 \sqrt[3]{\frac{T}{s}} = 1.72 \sqrt[3]{\frac{10,400}{8000}} = 1.72 \sqrt[3]{1.30} \\
&= 1.72 \times 1.09 \\
&= 1.88
\end{aligned}
\tag{13}
$$

Use $1^{15}/_{16}$ in. diam.

Illustrative Example 2. A shaft must resist a bending moment of 14,500 lb-in. and a torque of 11,800 lb-in. Calculate the ideal twisting moment.

$$
\begin{aligned}
T_i &= M + \sqrt{M^2 + T^2} = 14,500 + \sqrt{(14,500)^2 + (11,800)^2} \\
&= 14,500 + \sqrt{210,000,000 + 139,000,000} \\
&= 14,500 + \sqrt{349,000,000} = 14,500 + 18,700 \\
&= 33,200 \text{ lb-in.}
\end{aligned}
\tag{21}
$$

Illustrative Example 3. Investigate the safety of a $1\frac{1}{2}$-in.-diam shaft which must resist a bending moment of 4200 lb-in. and a torque of 2750 lb-in.   Use the method of the equivalent bending moment and a value of the design stress for bending of 12,000 psi.

$$M_e = \frac{1}{2}(M + \sqrt{M^2 + T^2}) = \frac{1}{2}[4200 + \sqrt{(4200)^2 + (2750)^2}]$$
$$= \frac{1}{2}(4200 + \sqrt{17,600,000 + 7,560,000})$$
$$= \frac{1}{2}(4200 + \sqrt{25,160,000}) = \frac{1}{2}(4200 + 5020)$$
$$= \frac{1}{2}(9220) = 4610 \text{ lb-in.} \tag{22}$$

The value of I for a solid circular section is given in the table on page 215 as $\pi d^4/64$ and $c = d/2$.   Hence,

$$\frac{I}{c} = \frac{\pi d^4}{64} \times \frac{2}{d} = \frac{\pi d^3}{32}$$
$$\frac{M}{s} = \frac{I}{c} \tag{23}$$

By substituting M_e for M,

$$M_e = s \times \frac{I}{c} = s \times \frac{\pi d^3}{32} = \frac{12,000 \times 3.14 \times 1.5 \times 1.5 \times 1.5}{32}$$
$$= 3980 \text{ lb-in.}$$

The shaft diameter is too small because it can resist safely no more than a bending moment of 3980 lb-in., whereas it is required to resist 4610 lb-in.   This problem might also have been solved by substituting the value of 4610 lb-in. for M_e in the flexure formula, calculating s, and finally comparing this calculated value with the given value of the design stress.

PROBLEMS

1. Calculate the values of the reactions (R_1 and R_2) on a simple beam 38 in. long. There is a concentrated load of 425 lb at a distance of 15 in. from R_1.   Neglect the weight of the beam.

2. A simply supported shaft beam is 52 in. in length and supports two concentrated gear loads, one of 580 lb at a distance of 21 in. from the left reaction, the other of 740 lb at 14 in. from the right reaction.   Neglecting the weight of the shaft, calculate the values of the reactions.

3. A simple beam is 8 ft long and is used to support a concentrated load of 600 lb and one of 850 lb located at 3 ft and 4.5 ft from the left end, respectively. In addition there is a uniform load of 100 lb per ft (including the weight of the beam) throughout the entire span.   Calculate the reactions.

4. A 14-in.-diam gear on a $2\frac{1}{2}$-in.-diam shaft is used to transmit a torque of 3150 lb-in.   The gear weighs 50 lb.   Calculate the maximum force of the gear.

5. Calculate the total force of a 12-in.-diam pulley weighing 88 lb and delivering 15 hp to a belt at 1260 rpm.

6. Calculate the maximum bending moment for a simply supported shaft beam $3\frac{1}{2}$ in. in diameter and 82 in. long.   A 150-lb gear with a tooth thrust of 370 lb is located 35 in. from the left support.   Assume that steel weighs 0.28 lb per cu in.

7. A gear 16 in. in diameter weighing 75 lb and a pulley weighing 120 lb are fixed to a simply supported shaft weighing 30 lb per linear foot. The shaft is 6 ft long and the gear and pulley are 1.5 ft and 4.0 ft from R_1, respectively. A torque of 2200 lb-in. is transmitted by the gear and the belt pulls on the pulley are 200 lb and 80 lb. Calculate:

a. The values of the reactions

b. The point of zero shear

c. The maximum bending moment

8. A shaft transmits 55 hp by means of a pulley at 200 rpm. The bending moment is 12,500 lb-in. Calculate the equivalent torque.

9. Design a solid circular shaft to meet the following conditions. The shaft is to be 7 ft long and supported at each end. A 250-lb rotor, 30 in. in diameter, is to be located 2.5 ft from the left end of the shaft. The tangential force on the rotor is 160 lb downward. The design stress in shear is 8000 psi and in bending 15,000 psi. Use the method of the ideal torque. Neglect the weight of shaft.

10. Investigate a 2-in.-diam solid circular shaft by the method of the equivalent bending moment to determine whether it can resist safely a twisting moment caused by a tooth thrust of 540 lb on a 20-in.-diam gear and a bending moment of 4500 lb-in. The design stress in bending is 14,000 psi.

11. A line shaft is required to transmit 60 horsepower at 110 rpm. The bending moment, caused by the weight of pulleys and belt pulls, is three-fourths of the twisting moment. Calculate the diameter for the most suitable standard size solid steel shaft. Assume the design stress for steel in shear as 7000 psi and use the method of the equivalent torque.

12. A solid steel shaft supported between bearings 60 in. apart transmits 150 hp at 400 rpm. There are two pulleys keyed to the shaft between bearings. The first is located 15 in. from the left support and its total force on the shaft is 1200 lb. The other pulley is located 25 in. from the right support and its total force is 2000 lb. Calculate the diameter of the shaft required. Use the method of the equivalent torque and a design stress for shear of 7500 psi. Neglect shaft weight.

13. Redesign the shaft of Prob. 10 for a hollow section with a ratio of $d/d_1 = 2$. This time use the method of the ideal torque and a design stress of 10,000 psi.

14. By the method of the equivalent torque, investigate a hollow shaft 3 in. OD and 1½ in. ID to determine whether it can be safely used to resist a twisting moment of 30,000 lb-in. and a bending moment of 24,000 lb-in. Use a design stress of 8000 psi.

15. A steel shaft transmits 1500 hp at 150 rpm. It must resist a bending moment equal to five-eighths of the twisting moment. Calculate the following items:

a. The outside diameter of this shaft when designed as a hollow section with a ratio of 3:1 outside to inside diameters

b. The fiber stress of a solid shaft whose diameter is the same as the outside diameter of the hollow section found in (a)

Use the method of the equivalent torque and a design stress of 10,000 psi.

16. Assuming values of 2000 lb-in. for the twisting moment and 3000 lb-in. for the bending moment, make calculations to compare the values of the stresses obtained by the use of the ideal torque method and the equivalent bending moment method. Assume also a convenient size of shaft such as 1 in. in diameter.

CHAPTER 10

ELASTIC DEFORMATION IN SHAFTS

52. Modulus of Rigidity. The stress induced in the steel bar of Fig. 56 (as discussed in Sec. 8) is found by dividing the load by the cross-sectional area in square inches, namely,

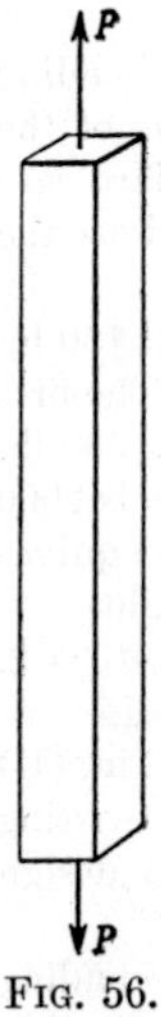

$$s = \frac{P}{A} \qquad (1)$$

Similarly, the amount of deformation for every inch of length (strain, δ) is found by dividing the total deformation Δ by the original length l of the bar in inches.

$$\delta = \frac{\Delta}{l} \qquad (2)$$

For deformations up to the elastic limit, the deformations are not permanent and the steel will return to its original size when the loads are removed. Deformations of this type are known as *elastic deformations*. Any stress and the corresponding value of the strain up to the elastic limit can be used for calculating the modulus of elasticity E by

Fig. 56.

$$E = \frac{s}{\delta} \qquad (3)$$

Modulus of elasticity values for steel in tension and compression vary between 28,000,000 and 30,000,000, depending on the type of steel.

From the study of strength of materials we will further recall that the concept of modulus of elasticity is also useful in cases of shear. Since torsion is in reality a shearing stress, the shear modulus of elasticity, or *modulus of rigidity* G, as it is more appropriately called, applies to shafts. The value of G is again the ratio of stress to strain,

$$G = \frac{s}{\delta} \qquad (24)$$

for values of s and δ at or below the elastic limit. As modulus of elasticity measures the stiffness of a material in tension or compression, modulus of rigidity measures the stiffness in shear. Its value for steel is around 12,000,000 or 13,000,000.[1] This shows that steel is more than twice as stiff in tension or compression as it is in torsion.

53. Torsional Deformation of Shafts. One end of a shaft of length l shown in Fig. 57 is assumed to be firmly fastened by the flange. A straight line BA is drawn on the surface parallel to the shaft axis and point B and the center Q connected. The free end is then subjected to a twisting moment as, for example, the force P_1 applied at the circumference of a gear. Because of the twisting, line BA will no longer

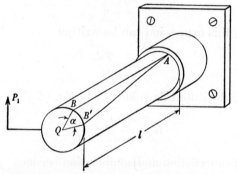

Fig. 57. Loads causing torsional deflection.

be straight and will begin to wind around the circumference, describing a *helix*. Point B will have moved to B'. Similarly, every point on the original line BA will have moved to a corresponding point on line $B'A$. The central angle BQB' is denoted by the symbol α (alpha) and is called the *angle of twist* or the angle of *torsional deflection*.

The length of the arc BB' is the *total* linear deformation of the shaft owing to torsional stress. This distance can be determined by means of a simple proportion, because BB' bears the same relationship to the entire circumference as α bears to 360°.

$$\frac{BB'}{\pi \times d} = \frac{\alpha}{360}$$

and

$$BB' = \frac{\alpha \times \pi \times d}{360}$$

The amount of linear deformation for each inch of length is the strain,

[1] As in the case of modulus of elasticity, the units for modulus of rigidity (pounds per square inch) are omitted in this text.

δ. Its value is obtained by dividing the value of BB' by the total length l in inches, namely,

$$\delta = \frac{\alpha \times \pi \times d}{360 \times l}$$

By substitution in $G = s/\delta$, we obtain

$$G = \frac{360sl}{\alpha \pi d} \tag{25}$$

The value of the stress as given in the basic torsion formula is

$$s = \frac{Tc}{J}$$

Since $c = d/2$, this expression can be written

$$s = \frac{Td}{2J}$$

Now we shall substitute this value of s in Eq. (25)

$$G = \frac{360l}{\alpha \pi d} \times \frac{Td}{2J}$$

which through cancellation and combination becomes

$$G = \frac{57.3Tl}{\alpha J} \quad \text{and} \quad \alpha = \frac{57.3Tl}{GJ} \tag{26}$$

This expression can be further modified to get the torsional deflection in terms of the diameter, since for a solid shaft $J = \pi d^4/32$,

$$\alpha = \frac{57.3Tl}{G} \times \frac{32}{\pi d^4}$$

By combining and simplifying, we obtain

$$\alpha = \frac{584Tl}{Gd^4} \tag{27}$$

An expression similar to Eq. (27) can also be developed for hollow shafts by substituting s from Eq. (15)

$$s = \frac{5.10T}{d^3(1 - q^4)}$$

in Eq. (25) to obtain

$$\alpha = \frac{584Tl}{Gd^4(1 - q^4)} \tag{28}$$

54. Limiting Torsional Deflection. A complete investigation or design of a beam calls for checking the maximum deflection. A beam may be sufficiently strong in bending, but because of excessive deflection, which would cause undue vibration, it cannot be used for certain purposes and a more rigid section must be chosen. In shafts, torsional deflection resulting in torsional vibration plays an even greater part than does deflection in beams. Limiting the torsional deflection is especially important in cases where the mechanism is highly synchronized. Remember too (Sec. 52) that steel is less than half as stiff when torsional stress is induced than under bending stress, composed as it is of tension and compression. Widely used specifications for shafts state that the angle of twist must not exceed 1° *for a length of shaft equal to* 20 *times the diameter.* For special-purpose shafting, the allowable torsional deformation is often much less.

In cases where limiting the torsional deflection is desirable, the application of this rule will often govern the value of the diameter. In other words, a diameter of shaft selected to be within the allowable design stress may deflect more than the allowable amount and a larger diameter shaft must be selected. In such cases it is quite proper to design the shaft by solving for d in Eq. (27) or Eq. (28). Let l equal $20d$ and $\alpha = 1°$.

Illustrative Example 1. A 3-in.-diam shaft of solid steel, 6 ft long, is used to resist a torque of 50,000 lb-in.

a. Investigate to determine whether the shaft is safe in torsion.

b. Investigate to determine whether the shaft is within allowable limits of torsional deflection.

c. If necessary, make calculations for the selection of a more suitable size shaft.

Use 10,000 psi for the design stress and 12,000,000 for the modulus of rigidity.

a.

$$s = \frac{16T}{\pi d^3} = \frac{5.10T}{d^3} = \frac{5.10 \times 50,000}{3 \times 3 \times 3} = 9450 \text{ psi} \quad (12)$$

The shaft is safe in torsion.

b.

Assume a length of $20d$, that is, 20×3.

$$\alpha = \frac{584Tl}{Gd^4} = \frac{584 \times 50,000 \times 20 \times 3}{12,000,000 \times 3 \times 3 \times 3 \times 3} = 1.8° \quad (27)$$

Since the shaft deflects more than 1° in a length equal to 20 diameters, there is excessive torsional deflection.

c.

$$\alpha = \frac{584 \times T \times 20d}{Gd^{43}} \tag{27}$$

$$d = \sqrt[3]{\frac{584 \times T \times 20}{G \times \alpha}} = \sqrt[3]{\frac{584 \times 50,000 \times 20}{12,000,000 \times 1}} = \sqrt[3]{48.6} = 3.65 \text{ in.}$$

Use $3\frac{3}{4}$ in. as the nearest standard size.

Illustrative Example 2. A hollow steel shaft must resist a torque of 128,000 lb-in. The torsional deflection is limited to 1° for a length of 20 times the outside diameter. The ratio of outside diameter to inside diameter is 2:1. Calculate the necessary diameters so that the allowable torsional deflection is not exceeded. Use 12,000,000 for the modulus of rigidity.

$$\alpha = \frac{584 \times T \times 20}{G \times d^3(1 - q^4)} \tag{28}$$

$$d = \sqrt[3]{\frac{584 \times T \times 20}{G \times \alpha(1 - q^4)}} = \sqrt[3]{\frac{584 \times 128,000 \times 20}{12,000,000 \times 0.938}} = \sqrt[3]{133} = 5.10 \text{ in.}$$

Use $5\frac{1}{2}$ in. as the next larger size for d and $5.5/2 = 2\frac{3}{4}$ in. for d_1.

55. Transverse Deflection in Shafts. Shafts, which are subjected to transverse loading to the extent that the bending stress must be considered, should also be investigated for the deflection caused by the bending (transverse deflection). One effect of a large transverse deflection is an excessive amount of wear on the bearings.

The general formula for transverse deflection Δ in beams is

$$\Delta = \text{a constant} \times \frac{Pl^3{}^*}{EI}$$

Special formulas to fit various manners of load distribution and various supporting conditions are available in engineering handbooks. Some of these are fairly complex.

In this text we shall discuss only the condition of one concentrated load on a simply supported span. The formula for the maximum deflection when the load is at *any point* in the span is

$$\Delta = \frac{Pc'}{3EIl} \sqrt{\left(\frac{cl + cc'}{3}\right)^3} \tag{29}$$

---

* Note the similarity between this formula and that discussed in the previous section for torsional deflection, Eq. (26), namely, $\alpha = 57.3Tl/GJ$.

Distances c and c' are as shown in Fig. 58. The length l is usually considered as the distance from center to center of bearings for deflection as well as bending moment calculations. Also l, c, and c' must be in inches.

When the concentrated load is at the mid-point of the span, the above expression reduces to

$$\Delta = \frac{Pl^3}{48EI} \qquad (30)$$

In general cases of shafting, a limit of 0.01 in. per ft of shaft (regardless of shaft diameter) for transverse deflection is frequently used. For high-speed shafting, the ordinary beam formulas do not apply because the centrifugal effect of the load adds to the deflection.

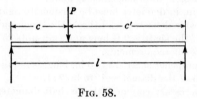

FIG. 58.

Illustrative Example. Calculate the maximum deflection of a 1½-in.-diam shaft beam 3 ft 8 in. between centers of bearings when a concentrated load of 200 lb is applied 1 ft 2 in. from one end. Is this deflection within allowable limits of 0.01 in. per ft of length?

$l = 3$ ft 8 in. $= 44$ in.

$c = 1$ ft 2 in. $= 14$ in.

$c' = 44 - 14 = 30$ in.

$$\Delta = \frac{Pc'}{3EIl} \sqrt{\left(\frac{cl + cc'}{3}\right)^3} \qquad (29)$$

$$\sqrt{\left(\frac{cl + cc'}{3}\right)^3} = \sqrt{\left(\frac{14 \times 44 + 14 \times 30}{3}\right)^3} = 6400$$

$$\Delta = \frac{Pc'}{3EIl} \sqrt{\left(\frac{cl + cc'}{3}\right)^3}$$

$$= \frac{200 \times 30 \times 64 \times 6400}{3 \times 30,000,000 \times 3.14 \times 1.5 \times 1.5 \times 1.5 \times 1.5 \times 44}$$

$= 0.0390$ in.

$$\text{Deflection per foot} = \frac{0.0390}{3.67} = 0.0106 \text{ in.}$$

This is slightly more than the allowable deflection of 0.01 in. per ft.

PROBLEMS

Unless otherwise specified a value of 12,000,000 should be assumed for the modulus of rigidity and 30,000,000 for the modulus of elasticity of steel.

1. A shaft of $1\frac{5}{8}$ in. in diameter shows an angular torsional deformation of 3° in a length of 15 in. when stressed to its elastic limit of 31,500 psi. Calculate the modulus of rigidity of the material.

2. A 16-in.-diam pulley on a 2-in.-diam steel shaft, 32 in. long, drives a belt by a force of 2000 lb. Calculate the angle of twist.

3. Calculate the torque that would cause an angle of twist of 4.5° in a hollow steel shaft 6 ft long. The outside diameter is 3 in. and $d/d_1 = 3$. Assume the modulus of rigidity as 13,000,000.

4. Investigate to determine whether a $2\frac{1}{2}$-in.-diam solid steel shaft, 6 ft 6 in. long, is within allowable limits as to both stress and torsional deflection when resisting a torque of 18,500 lb-in. Use a design stress of 8000 psi for torsion and a limiting angle of twist of 1° for a length of 20 diam.

5. A steel shaft 2 in. in diameter must be torsionally rigid within $\frac{1}{4}$° for every 3 ft of length. A greater deflection will throw the synchronization of the mechanism out of phase. If the shaft is 9 ft long and transmits 25 hp at 750 rpm, determine whether it meets the requirements.

6. A hollow steel propeller shaft, 15 ft 6 in. long, is required to transmit 90 hp at 70 rpm. The ratio of the diameters is to be 2 : 1.

a. Calculate the necessary commercial size shaft diameters for a design stress of 6000 psi.

b. Calculate the degrees of torsional deflection.

c. Determine whether this deflection is within allowable limits of 1° in 50 in.

7. A solid steel line shaft, 7 ft 4 in. long, must resist a torque of 35,000 lb-in. The limit stress is 24,000 psi and a factor of safety of 3 is necessary.

a. Calculate the most suitable shaft diameter to resist this torque.

b. Determine whether the torsional deflection for this diameter is within limits of 1° in 20 diam.

c. Recalculate the diameter if necessary.

8. Calculate the diameter for a solid steel shaft that will twist no more than is allowable when resisting a torque of 55,000 lb-in.

9. A solid steel shaft is required to transmit 25 hp at 500 rpm. The stress of 7000 psi and the torsional deflection of 1° for 20 diam are not to be exceeded. Calculate the proper size.

10. A 2-in. shaft is supported by bearings 4 ft 2 in. center to center. At the mid-point is a 14-in.-diam pulley with a net belt pull of 1600 lb.

a. Calculate the stress induced by this pull, when torsion only is considered.

b. How much will the shaft deflect transversely, if the resultant downward load is 2500 lb? Is this deflection within the safe limits of 0.01 in. per linear ft? Assume the modulus of elasticity as 29,000,000.

11. A $1\frac{1}{2}$-in.-diam shaft, 30 in. long, supports a spur gear at a distance of 20 in. from the left end. The shaft must not deflect transversely more than 0.01 in. for each foot of length. Calculate the maximum total load that may be caused by the gear.

CHAPTER 11

KEYS

56. Keys and Keyways. A key is a device for preventing motion between two adjacent machine parts, as, for example, between a pulley or gear and its shaft. Figure 59a illustrates a key fitted to a shaft with t, b, and l as the thickness, width, and length of the key, respectively.

A groove, called a *keyway* or *key seat*, is cut usually into both the shaft and the hub of the gear or pulley to accommodate the key. Not only is the shaft weakened in making the keyway because of the

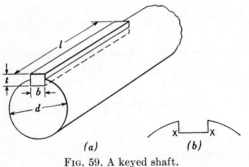

(a)   (b)

Fig. 59. A keyed shaft.

reduction in metal near the circumference where it is most needed, but also because of stress concentration. The inside (reentrant) corners of such a shaft are the points of greatest stress concentration, as indicated by the crosses in Fig. 59b. In view of these facts the key and keyway should not be made larger than good design proportions necessitate. The afore-mentioned Code for the Design of Transmission Shafting, published by the ASME, stipulates a reduction of 25 per cent in the design stress for shafts with keyways.

There are times when the key is purposely designed to be less strong than the shaft. If failure occurs under these circumstances, the key breaks rather than the shaft, thereby keeping the cost of replacement to a minimum.

Cold-rolled bar stock is frequently used for keys and the key seats milled to fit. The composition is medium plain carbon steel (around AISI 1035). Sometimes alloy steels are used for extra strength.

57. Square Keys and Flat Keys. Figure 60 shows two ways in which a key might fail if it were not strong enough to transmit the torque between shaft and gear or pulley. In each instance the key was not able to resist the push from the shaft tending to drive the key counterclockwise and the force from the pulley holding back this

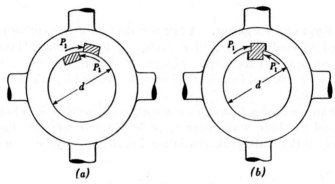

(a)   **(b)**

Fig. 60. Possible key failures.

pushing force from the shaft. These equal and opposite forces P_1 are assumed to be the same as the force at the shaft circumference, that is,

$$P_1 = \frac{T}{r}$$

where again T is the torque and r the radius of the shaft.

In Fig. 60a the portion of the key in the hub has slid past the portion in the shaft along the surface indicated—a typical shearing failure. The sheared area, $b \times l$, was stressed beyond the ultimate strength of the steel. Therefore, to resist safely the shearing forces P_1, the area must be of such size that it is not stressed beyond the design stress of the steel in shear, indicated here by s_s. In symbols,

$$P_1 = bls_s \quad \text{and} \quad l = \frac{P_1}{bs_s} \tag{31}$$

Let us now examine the second possible method of failure. In Fig. 60b the key again was too weak to resist successfully the push of the equal and opposite forces of the shaft and the hub, but this time the key was crushed between these forces. This is a type of compression failure known as bearing, analogous to bearing failure in riveted joints. The area of the key in the shaft on which the force P_1 was pushing and the area of the key in the hub on which the equal force P_1 was pushing was stressed beyond the ultimate bearing strength of the steel.[1] Therefore, to resist safely the forces P_1, the area in question must be of such size that it is not stressed beyond the *design* stress of the steel in bearing, s_b. Also, for the two areas to be equal, the key must extend into the hub the same distance that it extends into the shaft, which is $t/2$. In symbols,

$$P_1 = \frac{t}{2} \times l \times s_b$$

and

$$P_1 = \frac{tls_b}{2} \quad \text{also} \quad l = \frac{2P_1}{ts_b} \tag{32}$$

The force P_1 in the shearing investigation is the same as P_1 in the case of bearing; hence, by equating (31) and (32),

$$bls_s = \frac{tls_b}{2}$$

which by transposing and canceling becomes

$$\frac{2b}{t} = \frac{s_b}{s_s}$$

Let us assume that design stresses of 20,000 psi for bearing and 10,000 psi for shear are chosen or that, in any event, the ratio of the two is 2:1. By substitution,

$$\frac{2b}{t} = \frac{20,000}{10,000}$$

which reduces to

$$b = t$$

With these design stress assumptions, the best key design results in a key that is just as thick as it is wide—in other words, a *square* key. However, the design stress for bearing might very well be assumed as more than double the design stress for shear, in which case t may be made smaller than b to give a *flat* key, also known as a *rectangular* key.

---

[1] It is possible that a bearing failure would occur in the shaft and hub in addition to or instead of the bearing failure of the key. In any event, since the bearing areas are equal, the analysis does not change.

The accompanying table is taken from the American Standards Association table for square and flat keys. It will be noted that in each case the width b is approximately one-fourth of the shaft diameter. This ratio applies to other types of keys also.

TABLE 4. SQUARE AND FLAT KEYS

Shaft diameters	Square keys b and t	Flat keys	
		b	t
$1\frac{5}{16}$ –$1\frac{3}{8}$	$\frac{5}{16}$	$\frac{5}{16}$	$\frac{1}{4}$
$1\frac{7}{16}$ –$1\frac{3}{4}$	$\frac{3}{8}$	$\frac{3}{8}$	$\frac{1}{4}$
$1\frac{13}{16}$–$2\frac{1}{4}$	$\frac{1}{2}$	$\frac{1}{2}$	$\frac{3}{8}$
$2\frac{5}{16}$ –$2\frac{3}{4}$	$\frac{5}{8}$	$\frac{5}{8}$	$\frac{7}{16}$
$2\frac{7}{8}$ –$3\frac{1}{4}$	$\frac{3}{4}$	$\frac{3}{4}$	$\frac{1}{2}$
$3\frac{3}{8}$ –$3\frac{3}{4}$	$\frac{7}{8}$	$\frac{7}{8}$	$\frac{5}{8}$
$3\frac{7}{8}$ –$4\frac{1}{2}$	1	1	$\frac{3}{4}$
$4\frac{3}{4}$ –$5\frac{1}{2}$	$1\frac{1}{4}$	$1\frac{1}{4}$	$\frac{7}{8}$
$5\frac{3}{4}$ –6	$1\frac{1}{2}$	$1\frac{1}{2}$	1

In design, the ratio of the design stress in bearing to that in shear will indicate whether a square or flat key is to be chosen. Select the proper size from Table 4, and calculate the necessary length.

Illustrative Example 1. A $2\frac{1}{2}$-in.-diam shaft resists a torque of 25,000 lb-in. Assume design stresses of 10,000 psi in shear and 20,000 psi in compression (bearing) to select the proper type and size of key from the foregoing table. Calculate the required minimum length of key.

Because of the assumption that s_s is $\frac{1}{2}s_b$, a square key should be chosen. The table gives a $\frac{5}{8}$- by $\frac{5}{8}$-in. key for a $2\frac{1}{2}$-in.-diam shaft.

$$P_1 = \frac{T}{r} = \frac{25,000}{1.25} = 20,000 \text{ lb}$$

In shear:

$$l = \frac{P_1}{bs_s} = \frac{20,000}{0.625 \times 10,000} = 3.2 \text{ in.} \tag{31}$$

An investigation for bearing should result in an equal length.

$$l = \frac{2P_1}{ts_b} = \frac{2 \times 20,000}{0.625 \times 20,000} = 3.2 \text{ in.} \tag{32}$$

Illustrative Example 2. Select the proper cross-sectional dimensions and length of a key to resist a torque of 40,500 lb-in. on a 3-in.-diam shaft. Assume a design stress for bearing of 23,000 psi and for shear of 8000 psi.

This time a flat key should be selected. By reference to the table we see that a $\frac{3}{4}$- by $\frac{1}{2}$-in. key is recommended.

$$P_1 = \frac{T}{r} = \frac{40,500}{1.5} = 27,000 \text{ lb}$$

In shear:

$$l = \frac{P_1}{bs_s} = \frac{27,000}{0.75 \times 8000} = 4.50 \text{ in.} \qquad (31)$$

In bearing:

$$l = \frac{2P_1}{ts_b} = \frac{2 \times 27,000}{0.5 \times 23,000} = 4.70 \text{ in.} \qquad (32)$$

Choose the larger value to make key $4\frac{3}{4}$ in. long.

Square keys and flat keys are made both plain and tapered. The t dimension in the taper key is the one that varies. The taper helps in driving and removing and ensures a tight fit when in place. As the key is driven in, the opposite side of the shaft is forced tightly against the hub, causing a large frictional resistance to independent rotation of hub and shaft and consequently less force (P_1) on the key when the system is in operation.

58. Woodruff Keys. Woodruff keys (Fig. 61a and b) resemble segments of circles in shape. The shaft (Fig. 61b) is carefully milled to accommodate the circular portion and to allow the key to extend

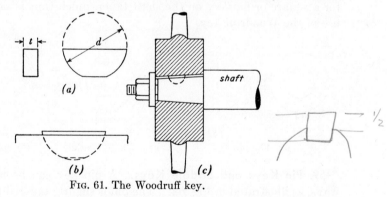

Fig. 61. The Woodruff key.

into the hub a distance of one-half the thickness of the key ($t/2$). An abbreviated table of Woodruff key numbers and sizes is given in Table 5. Note that the cutter diameter d, which is practically the same as the length dimension of the key, is always about four times the key thickness t. The proper size of key for a shaft is one in which the t dimension is about one-fifth of the shaft diameter. Two or more keys may be used if found necessary. Woodruff keys are suitable for light torques only.

TABLE 5. WOODRUFF KEYS

Key number	t (thickness)	d (cutter diam)
3	$\frac{1}{8}$	$\frac{1}{2}$
5	$\frac{1}{8}$	$\frac{5}{8}$
6	$\frac{5}{32}$	$\frac{5}{8}$
8	$\frac{5}{32}$	$\frac{3}{4}$
9	$\frac{3}{16}$	$\frac{3}{4}$
11	$\frac{3}{16}$	$\frac{7}{8}$
12	$\frac{7}{32}$	$\frac{7}{8}$
14	$\frac{7}{32}$	1
15	$\frac{1}{4}$	1
18	$\frac{1}{4}$	$1\frac{1}{8}$
C	$\frac{5}{16}$	$1\frac{1}{8}$
D	$\frac{5}{16}$	$1\frac{1}{4}$

Among the advantages of this type of key is the ease of removal from the shaft. Light taps with a hammer on a blunt cold chisel are sufficient for this operation. Figure 61c illustrates a situation where the Woodruff key is used to advantage. The tapered hub is secured to the end of the shaft, which is tapered to fit, by means of the stud bolt. Because of the difficulty of milling a uniform groove keyway for a square or flat key on the shaft taper, much time is saved by the use of the Woodruff key.

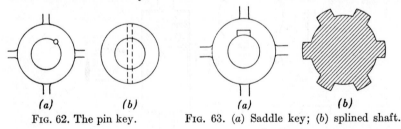

(a)   (b)   (a)   (b)
FIG. 62. The pin key.   FIG. 63. (a) Saddle key; (b) splined shaft.

59. Pin Keys and Saddle Keys. A pin key can be used in two ways, as illustrated in Fig. 62. The pin is slightly tapered to facilitate assembly. When used as in Fig. 62a, the pin key is analogous to a square key and should be equally as strong. The method of fitting the pin through the hub, perpendicular to the shaft (Fig. 62b) is suitable to very light torques only. It is used extensively for fastening collars on shafts.   Examine the areas resisting the shearing stresses in Fig. 62a and b to find out why the second assembly is incapable of heavy duty.   In Fig. 62b the hole is reamed to fit the taper of the pin.

No part of the saddle key of Fig. 63a protrudes into the shaft. The curved surface of the key in contact with the shaft circumference is of a slightly smaller radius than the shaft.   The key is tapered and, when driven into place, acts as a powerful compression spring, thus creating great frictional resistance between shaft and hub.   However, since there is no further means of holding the two together, the saddle key cannot resist very heavy torques.   The use of this type of key is advantageous when the position of the gear or pulley on the shaft is to be changed periodically.   Why?

60. Feather Keys and Splines.   One means of accomplishing the frequent engagement and disengagement of a gear is by the use of a *feather key*.   This type of key fits loosely between shaft and hub, allowing the gear to be moved parallel to the shaft to or away from the mating gear.   In some types the key is so constructed that it can be made fast to the hub and in other types it is fastened by countersunk screws to the shaft.

Spline fittings were developed by and are extensively used in the automobile industry.   They are made by cutting longitudinal grooves in a shaft with the milling machine to leave protrusions which form the keys.   One type is shown in Fig. 63b.   The hubs are made with a corresponding number of key seats.   Increasing the number of keys, of course, increases the key strength proportionately.   Splined shafts and hubs are machined either for permanent fitting or for sliding of hub along shaft.

PROBLEMS

1. A flat key $\frac{1}{2}$ in. wide, $\frac{3}{8}$ in. thick, and 3 in. long is required to transmit a torque of 10,500 lb-in. from a 2-in.-diam shaft.   Investigate to determine whether the length is sufficient.   Use a design stress in shear of 8000 psi and in bearing of 20,000 psi.

2. Select and check by calculation the proper dimensions for a square key to resist a net belt pull of 600 lb at the circumference of an 8-in.-diam pulley on a $1\frac{1}{2}$-in.-diam shaft.   Use design stresses of 10,000 psi and 20,000 psi for shear and bearing, respectively.

3. Calculate the shearing and bearing stresses on a square key, $\frac{7}{8}$ in. by 6 in. in length, when resisting a tangential force of 43,000 lb at the circumference of the shaft.

4. Select and check by calculation the proper dimensions for a rectangular key to fit a $3\frac{1}{2}$-in.-diam shaft which is stressed to 6000 psi at its extreme fibers.   The design stress in shear for the key is 9000 psi and in bearing is 22,000 psi.

5. A gear 26 in. in diameter is used to transmit 120 hp at 240 rpm on a 3-in.-diam shaft.   The length of the square key to be used is 5 in.   Calculate the cross-sectional dimension that will limit the shearing stress in the key to 7000 psi.

6. Select the proper key number for a Woodruff key to fit on a 1-in.-diam shaft. How many of these keys are needed when used to transmit 14 hp through a gear at 500 rpm?   The shearing stress in the key should not exceed 10,000 psi.

7. Calculate the necessary diameter for a wrist pin for a gasoline engine, if the piston diameter is $3\frac{1}{2}$ in. and the explosive pressure is 400 psi.   Assume the design stress in shear as 8000 psi.

8. A pin key of $\frac{3}{8}$-in. average diameter is driven parallel to the axis of a 2-in.-diam shaft.   Half of the pin protrudes into the shaft and half into the hub of a pulley.   Calculate the maximum torque that this pin can resist safely in shear and bearing per inch of length.   Use a value of 10,000 psi for the design stress in shear and 20,000 psi for that in bearing.   *Hint:* The area of hub and key resisting bearing stress is considered to be equal to that of a square key.

9. A propeller of an outboard motor has a $1\frac{1}{8}$-in.-OD hub which fits on a $\frac{5}{8}$-in.-diam shaft.   The hub and shaft are fastened by a brass shear pin.   If an overload occurs at the propeller, the pin will shear, thus avoiding damage to the rest of the mechanism.   Calculate the diameter of the shear pin which will fail at a torque of 600 lb-in. with a shearing stress of 40,000 psi.

CHAPTER 12

PULLEYS AND BELTS

61. Pulley Materials and Types. Pulleys are wheels used to transmit power from one shaft to another by means of belts. In Fig. 64 both driver and driven are rotating in the counterclockwise direction. A reversal of direction can be accomplished by crossing the belt. Because there is a certain amount of slippage between both

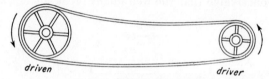

driven   driver

FIG. 64. Pulleys and belt.

driver and driven pulley and the belt, pulleys should not be used where an exact velocity ratio must be maintained.

The materials from which pulleys are made are forged steel, wood, compressed paper pulp, and, most common of all, cast iron. The principal parts of a pulley are the hub, adjacent to the shaft, the arms, extending radially from the hub, and the rim at the circumfer-

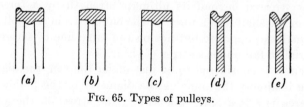

(a)   *(b)*   *(c)*   *(d)*   *(e)*

FIG. 65. Types of pulleys.

ence. Instead of arms, some pulleys have a solid piece between the hub and the rim. These are known as solid-web pulleys. Some pulleys are made in two semicircular halves and bolted or riveted together. These are called split pulleys.

The drawings of Fig. 65 illustrate four types of pulley rims: flanged, crowned, flat, and grooved, respectively. The flanged pulley (Fig.

65a) affords a positive means of keeping the belt on the pulley. It is used with vertical shafts. The crowned rim (Fig. 65b) is very generally used for larger pulleys. A moving belt tends to place itself centrally on a crowned pulley, thus continually counteracting any slight force tending to cause sidewise displacement. Flat pulleys (Fig. 65c) are used in cases where it is necessary to shift the belt from one pulley to another, as in a tight pulley (keyed to the shaft) and a loose pulley (not keyed) placed side by side. Two such pulleys act as a clutch to engage and disengage a mechanism. The grooved type of rim (Fig. 65d and e) is one in which the cross section is made either with a semicircular or V groove to accommodate a round or V belt. Belts and pulleys of this type are used for low power transmission and the pulleys are usually of the solid-web construction.

62. Design of Solid-web Pulley. Figure 66a shows a way in which a solid-web pulley might fail. If the web is not strong enough to hold under the driving force of the shaft and the resisting force of the pulley, it is conceivable that the web might break along some circular

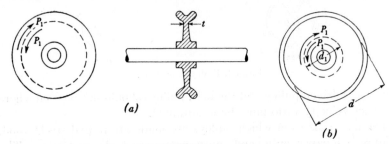

(a)

(b)

Fig. 66. Solid-web pulley.

line as the dotted circle shown. The surface represented by the dotted circle was stressed beyond its ultimate strength by a force on the inside from the shaft and a force on the outside from the belt. These forces, equal and opposite but not on the same line of action, caused the material to fail under a stress of shear.

If the circle is considered to be just under the rim, the diameter of this circle is almost the same as the diameter of the pulley and the sheared area is $\pi \times d \times t$. For safety, the stress on the area must not exceed the design stress of the material in shear. Also because the circle is very near the rim, the value of the shearing forces is practically the same as the net pull of the belt on the rim itself. This pull is P_1. In symbols,

$$P_1 = \pi dts \quad \text{and} \quad t = \frac{P_1}{\pi ds} \tag{33}$$

When a circle near the *hub* is examined, it will be found that this section calls for a considerably thicker web. Why is this so? In the first place, to obtain an equal area, t must be greater because d_1 is much smaller than d (Fig. 66b). Second, however, a *greater* area is also needed because of the fact that the value of the shearing forces increases as we approach the shaft. You can see that this is true because the *moment arm* has decreased while the torque remains constant. For these reasons, the web of a solid-web pulley is made thickest at the hub and thinnest at the rim. In design, the calculated thickness at the rim will usually come out very small. The designer then conforms to standard practice of minimum thicknesses. A good rule is never to make a cast-iron web less than $\frac{3}{8}$ in. or a steel web less than $\frac{1}{4}$ in. in thickness at the rim.

Illustrative Example. Calculate the minimum thickness of web required at the rim of a 5-in.-diam solid-web steel pulley which must deliver 5 hp at a speed of 800 rpm to a ventilating blower. Use a design stress of 7000 psi. What thickness should be selected?

$$T = \frac{63{,}000H}{n} = \frac{63{,}000 \times 5}{800} = 394 \text{ lb-in.} \tag{11}$$

$$P_1 = \frac{T}{r} = \frac{394}{2.5} = 158 \text{ lb}$$

$$t = \frac{P_1}{\pi ds} = \frac{158}{3.14 \times 5 \times 7000} = 0.00144 \text{ in.} \tag{33}$$

To conform with practice for minimum thickness at rim, make thickness $\frac{1}{4}$ in.

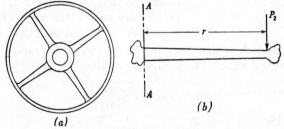

(a) *(b)*

FIG. 67. The arm pulley.

63. Design of Arm Pulley.

As in the case of the solid-web pulley, the arm pulley (Fig. 67) is stressed in shear when resisting the equal and opposite forces of shaft and belt. However, the designer regards each arm as a cantilever beam, fixed at the hub and supporting a concentrated load at the rim. As with most such beams, the arm should be safeguarded against a bending rather than a shear failure. The applied force which the arms must support is again the force P_1 equal to T/r.

At any time during rotation about half the total number of arms are connected to a portion of the rim which is *not* in contact with the belt. The best design practice is to consider one-half of all the arms as helping to resist the force P_1 at any one instant. For three- and four-arm pulleys, we assume two arms to be acting, while for five- and six-arm pulleys, we assume three arms. If P_2 is the concentrated load at the end of each cantilever acting (Fig. 67b), then

$$P_2 = \frac{P_1}{\frac{1}{2} \text{ number of arms}} \qquad (34)$$

The maximum bending moment will occur at the hub (section AA in Fig. 67b) and is P_2 multiplied by the length of the arm. In practice, unless a hub diameter is specified, the radius of the pulley is used for this distance. Therefore,

$$M_{\max} = P_2 \times r$$

The flexure formula, used in beam design,

$$\frac{M}{s} = \frac{I}{c} \qquad (23)$$

may then be applied. The terms are as previously explained (Sec. 51).

Figure 68 shows an ellipse with center-of-gravity axes 1-1 and 2-2. Arms of pulleys are most often made elliptical in cross section, placed with the long axis of the ellipse parallel to the plane of the pulley. This means that axis 1-1 is the neutral axis. For a circular section,

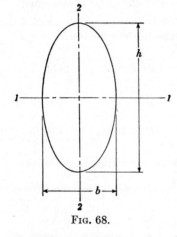

FIG. 68.

$$I_{1-1} = I_{2-2} = \frac{\pi d^4}{64}$$

and for an elliptical section,

$$I_{1-1} = \frac{\pi b h^3}{64}$$

as given in the table on page 215.

Now let us calculate the value of I/c, the section modulus, for an elliptical section.

$$\frac{I}{c} = \frac{\pi b h^3}{64} \times \frac{2}{h} = \frac{\pi b h^2}{32}$$

which can be simplified to

$$\frac{I}{c} = 0.0981bh^2$$

In machine-design practice, the major axis h of the ellipse is usually made either twice or three times as long as the minor axis b, that is, for a ratio of 2:1, $h = 2b$ and $h^2 = 4b^2$. Therefore,

$$\frac{I}{c} = 0.0981 \times b \times 4b^2 = 0.392b^3$$

and, since $M/s = I/c$,

$$\frac{M}{s} = 0.392b^3$$

Then, by solving for b,

$$b = \sqrt[3]{\frac{M}{0.392s}} \qquad (35)$$

The same steps are followed when the ratio of h to b is 3:1. In this case,

$$h = 3b \qquad \text{and} \qquad h^2 = 9b^2$$

$$\frac{I}{c} = 0.0981 \times b \times 9b^2 = 0.883b^3$$

$$\frac{M}{s} = 0.883b^3$$

$$b = \sqrt[3]{\frac{M}{0.883s}} \qquad (36)$$

Under the assumption that the arms act as cantilever beams,[1] there is no bending moment and, hence, no bending stress at the rim end, although shearing stress is present throughout the length. For economy, therefore, and appearance, the arms are tapered from hub to rim; the taper is $\frac{1}{4}$ to $\frac{3}{8}$ in. per ft. These values give ample area on the outer end to resist the shearing forces.

Illustrative Example. A 60-in.-diam five-arm cast-iron pulley is to deliver 300 hp from the shaft of a water turbine at a speed of 150 rpm. The arms are elliptical in cross section, with the major and minor axes in the ratio of 2:1. Calculate the dimensions of each axis at the hub end of the arm. Assume a design stress value of 2500 psi for cast iron in bending.

[1] This assumption is not strictly according to fact, but is adhered to in this text because it is in conformity with design practice. It would appear that an analysis based on the assumption that the arm is a beam fixed at both ends would yield more accurate results.

$$T = \frac{63,000H}{n} = \frac{63,000 \times 300}{150} = 126,000 \text{ lb-in.} \quad (11)$$

$$P_1 = \frac{T}{r} = \frac{126,000}{30} = 4200 \text{ lb}$$

The number of arms considered to be acting for a five-arm pulley is three.

Force at end of each arm: $P_2 = {}^{4200}\!/_3 = 1400$ lb.

$$M_{max} = P_2 \times r = 1400 \times 30 = 42,000 \text{ lb-in.}$$

$$b = \sqrt[3]{\frac{M}{0.392s}} = \sqrt[3]{\frac{42,000}{0.392 \times 2500}} = \sqrt[3]{42.8} = 3.50 \text{ in.} \quad (35)$$

$$h = 2b = 7.00 \text{ in.}$$

64. Selecting the Proper Pulley. To select a pulley for a particular use, as previously stated, an engineer most frequently makes a choice that will best suit his needs from a manufacturer's catalogue. In so doing, the following considerations are important.

1. *Pulley diameter.* The diameter must be such as to provide the proper velocity ratio, that is, to develop the desired rpm in the machine or countershaft to which the power is being delivered.

2. *Hub length.* The hub must be as least as long as the calculated length of the key. Why?

3. *Rim type.* One type of rim is best for the particular conditions of operation.

4. *Rim width.* The width of rim (width of pulley face) should be about 25 per cent more than the width of the belt.

5. *Strength.* Finally, it is advisable to check the strength of a tentative selection according to Secs. 62 or 63.

65. Belting—Materials and Splices. The oldest and still the most widely used material from which belts are made is *leather*. Since cowhide averages about $\frac{3}{16}$ in. in thickness, thicker belts are made by cementing the hides to form layers. Thus we have the terms single, double, or triple belts. *Rubber* is less strong than leather for belting but is harmed less by moisture. The rubber is held by a matrix of heavy canvas and the belts are built up with varying numbers of plies of canvas and rubber to as many as 15. *Balata* belts are made in a similar way to rubber belts with balata, a gummy substance, substituted for rubber.

A belt is no stronger than its joint and, therefore, in the interest of economy, particular attention should be paid to the manner of splicing the ends. The most effective splice is a cemented lap joint. The ends of the belt are first cut to a chisel edge to avoid a lump, and then the joint is cemented and held together under pressure. Butt joints held by wire links or rawhide lacing are less strong.

66. Belt Design. Pulleys and belts depend for successful operation on the frictional resistance between the surfaces of these two parts. The greater the coefficient of friction, the less belt tension is needed, and, hence, the smaller the necessary belt size. Also less belt tension means lower frictional losses and less wear on the shaft bearings. One added advantage of leather belting over rubber is the greater friction between leather and the pulley surface, whether that surface is metal, wood, or fiber.

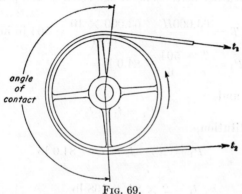

Fig. 69.

Design stresses in tension for leather range from about 250 to 450 psi. However, for any particular belt, a chosen value must be reduced according to the strength of the joint. The percentage of a joint's strength in relation to that of the solid material is known as the *efficiency* of the joint. As a formula, where η (Greek letter *eta*) stands for efficiency,

$$\eta = \frac{\text{strength of joint}}{\text{strength of solid material}} \tag{37}$$

If the joint has an efficiency of, let us say, 70 per cent, then the design stress for the belt should be reduced accordingly.

The centrifugal effect of all belts in motion causes an increase in the belt tension. In low-speed belting this factor is usually neglected but should be considered for high speeds.

The theory of belt design is quite complicated, involving such factors as the angle of contact (between belt and pulley, see Fig. 69) and the slip of the belt on the pulley, as well as the frictional resistance and the centrifugal force already mentioned. In this text a more simple procedure will be presented based on certain assumptions. For example, the given design stress will be assumed to have already been reduced for the centrifugal tendency where necessary. When the greatest belt tension (tension on the tight side) is known, then the area of the belt cross section can readily be calculated from

$$A = \frac{P}{s} \tag{1}$$

Illustrative Example. A 12-in.-diam pulley on a motor shaft transmits 10 hp at a speed of 1250 rpm. The ratio of tight to slack side of the belt is assumed to be 2:1. Calculate the necessary width of a $\frac{3}{16}$-in. single leather belt. The design stress is 200 psi.

$$T = \frac{63{,}000H}{n} = \frac{63{,}000 \times 10}{1250} = 504 \text{ lb-in.} \tag{11}$$

$$P_1 = \frac{T}{r} = \frac{504}{6} = 84.0 \text{ lb}$$

Since $t_1 = 2t_2$ and

$$P_1 = t_1 - t_2$$

then, by substitution,

$$P_1 = 2t_2 - t_2 = t_2 = 84.0 \text{ lb}$$

and

$$t_1 = 2 \times 84 = 168 \text{ lb}$$

This is the maximum tension that must be carried by the belt.

$$A = \frac{P}{s} = \frac{168}{200} = 0.84 \text{ sq in.} \tag{1}$$

$$= \frac{3}{16} \times \text{width}$$
$$\text{Width} = 0.84 \times 1\frac{6}{3} = 4.48, \text{ or } 4.50 \text{ in.}$$

When, however, values (or ratio) of the belt tensions are not given, the maximum belt tension can be calculated by the approximate method as follows.

Figure 70 shows a weight W of some kind being moved horizontally on a plane by force F'. The magnitude of F' is just enough to overcome the frictional resistance F and to keep W in uniform motion, that is, $F' = F$. Force N represents the push of the plane on W. Of course, N must be equal and opposite to W. Less force is necessary to pull the weight than to lift it, and therefore F is less than W or N.

The ratio of F to N is called the *coefficient of friction f* and, expressed mathematically,

$$f = \frac{F}{N} \tag{38}$$

We know that the value of the frictional resistance (for a given N) and hence the coefficient of friction varies with the type of materials in contact and the smoothness of the contact surfaces. The area of the contact surfaces has very little effect.

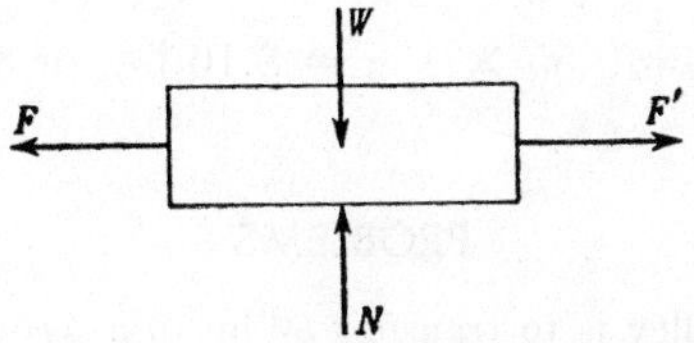

FIG. 70.

In the case of a belt and pulley (Fig. 69), let us assume that the angle of contact is 180°, which assumption is usually approximately correct. The force N pushing belt and pulley together will then be $t_1 + t_2$, or

$$N = t_1 + t_2$$

Also the maximum value of P_1 would be that value which would cause the belt to slip on the pulley. This is F' or F. By substitution of these values in $f = F/N$, Eq. (38), we have

$$f = \frac{P_1}{t_1 + t_2}$$

But $\quad t_2 = t_1 - P_1$ and, by substitution,

$$f = \frac{P_1}{t_1 + t_1 - P_1} = \frac{P_1}{2t_1 - P_1}$$

and

$$2t_1 f - P_1 f = P_1$$

also

$$t_1 = \frac{P_1 + P_1 f}{2f} \tag{39}$$

Illustrative Example. A 42-in.-diam cast-iron pulley transmits 96 hp at a speed of 800 rpm. Calculate the necessary width for the $\frac{5}{16}$-in. double leather belt. The value of the coefficient of friction between cast iron and leather is 0.40. Assume 330 psi as the safe stress for the belting and a joint efficiency of 75 per cent.

$$T = \frac{63,000H}{n} = \frac{63,000 \times 96}{800} = 7560 \text{ lb-in.} \tag{11}$$

$$P_1 = \frac{T}{r} = \frac{7560}{21} = 360 \text{ lb}$$

$$t_1 = \frac{P_1 + P_1 f}{2f} = \frac{360 + 360 \times 0.40}{2 \times 0.40} = \frac{504}{0.80} = 630 \text{ lb} \tag{39}$$

Strength of joint $= \eta \times$ strength of solid material   (37)

Strength of joint in psi $= 0.75 \times 330 = 248$

$$A = \frac{P}{s} = \frac{630}{248} = 2.55 \text{ sq in.} \tag{1}$$

Width $= 2.55 \times 1\%_5 = 8.16$ in., or $8\frac{1}{2}$ in.

PROBLEMS

1. A 30-in.-diam pulley is to transmit 50 hp to a second pulley with a speed ratio of 1:2.5 and a belt velocity of 1000 ft per min.   Calculate:

a. The diameter of the second pulley

b. The rpm of the second pulley

2. Calculate the velocity of a belt in feet per minute which transmits 72 hp from a 20-in.-diam pulley.   The net belt pull is 600 lb.

3. A 4-in.-diam cast-iron solid-web pulley has a web thickness of $\frac{3}{8}$ in. at the circumference and $\frac{1}{2}$ in. at the hub.   The hub diameter is $1\frac{1}{2}$ in.   Make calculations to determine whether the web thicknesses are sufficient for the pulley to transmit 3 hp at a speed of 200 rpm.   A design stress of 1800 psi for cast iron in shear is to be used for the conditions that obtain.

4. An 18-in.-diam four-arm steel pulley is used to transmit 30 hp at 550 rpm. Each arm is elliptical in cross section with axes dimensions of 1.5 and 3.0 in. at the hub.   Make calculations to determine whether the arms are of sufficient strength.   Use a design stress of 10,000 psi.

5. Calculate the dimensions of the major and minor axes (ratio of 3:1) of a 22-in. elliptical arm pulley with five arms.   The pulley is keyed to a $2\frac{1}{2}$-in.-diam shaft which revolves at 500 rpm and is stressed to 8000 psi.   Assume a design stress of 2100 psi for the cast-iron pulley.

6. A 2-hp motor, speed 1725 rpm, is belted to an air compressor, the speed of which is 400 rpm.   The diameter of the motor pulley is 4 in.   Calculate the axes lengths of the cross sections of the arms of the compressor pulley if there are to be six arms elliptical in cross section with the major axis twice that of the minor. The design stress is 2000 psi.

7. A cast-iron pulley 26 in. in diameter has a hub diameter of 4 in. and five arms rectangular in cross section.   The pull on the tight side of the belt is 260 lb and on the slack side is 140 lb.   Calculate the cross-sectional dimensions necessary for the arms, if a ratio of 3:1 is to be held.   Use a design stress of 2400 psi.

8. Investigate a belt $\frac{1}{4}$ in. by 6 in. in cross section to determine whether it can safely transmit 12 hp from a 20-in.-diam pulley which rotates at 400 rpm. Assume a ratio of 2:1 for tight to slack side and a design stress of 300 psi in tension for the belt.

9. How much horsepower can a $\frac{5}{16}$- by 8-in. double leather belt safely transmit when running over a 24-in.-diam pulley at 280 rpm. Use a ratio of 2.5:1 for tight to slack side and a design stress of 350 psi for the belt in tension.

10. A 10-in. motor pulley exerts a pull of 180 lb on the driving side of a belt and 100 lb on the return side. The belt drives a 32-in. pulley on a solid circular shaft. The motor speed is 1325 rpm.

    a. Calculate the value of the horsepower transmitted.

    b. Calculate the size of shaft required for the driven pulley. Design stress equals 8000 psi.

    c. If the design stress for the belt is 200 psi, calculate the width of a single belt, $\frac{3}{16}$ in. thick, suitable for this drive.

11. A 24-in.-diam pulley transmits 22 hp by means of a rubber belt $\frac{3}{8}$ in. thick when rotating at 380 rpm. Calculate the required belt width. The design stress for the rubber is 200 psi with a joint efficiency of 80 per cent. Use a value of 0.35 for the coefficient of friction between belt and pulley.

12. A torque of 15,000 lb-in. is resisted by a 36-in.-diam pulley. The belt is $\frac{1}{2}$ in. by 9 in. in cross section and the design stress of the leather is 400 psi with a joint efficiency of 85 per cent. What is the necessary coefficient of friction?

CHAPTER 13

GEARS AND FRICTION WHEELS

67. Uses and Types of Gears. As in the case of pulleys, gears are used to transmit power from one shaft to another. Figure 71 illustrates power transmission by use of two disks or flat cylinders. The successful operation of this system depends only on the frictional grip of the lateral surfaces of the disks at the points of contact. You can see that for anything other than very small torques the driving disk would not move the other at all, and there would be constant slippage.

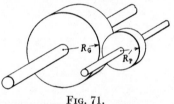

FIG. 71.

If, however, teeth were cut into the cylinders in such a way that each tooth on the driving cylinder would push a tooth on the driven cylinder, the slippage is eliminated and a positive means of power transmission afforded. Toothed wheels designed to mesh with other toothed wheels to transmit power are known as gears. Since there is no slippage, gears provide a means of maintaining an exact velocity ratio. For example, a velocity ratio of 2:1 is obtained by mating gears with twice as many teeth on the driven gear as there are on the driver.[1]

When the shafts are parallel as in Fig. 71, the teeth are formed on the lateral surfaces of low cylinders. Gears of this type are known as *spur* gears. Most of the time the teeth are made parallel to the shaft, but sometimes they are set at an angle and each tooth would describe a helix about the shaft if extended. These latter are called *helical* gears (commonly known as *spiral* gears).

When the shafts are not parallel (Fig. 72a), the teeth are formed on

[1] A sprocket is also a toothed wheel but transmits power through a chain to another sprocket. The teeth fit into the links of the chain. Sprockets and chains, therefore, are likewise similar to pulleys and belts, but, as in gears, slippage is eliminated.

the lateral surfaces of cone frustrums and are called *bevel* gears.   Bevel
gears in which the shafts are at an angle of 90° are *miter* gears.

A spur gear is also used to impart straight-line (rectilinear) motion,
as in Fig. 72*b*.   In this case the straight part is called a *rack* and the
gear the *pinion*.   Sometimes the rack is the driver, imparting a rotary
motion by means of its rectilinear motion.   The term pinion is also
used to describe the smaller of two mating gears, the larger being called
the gear.

For large reductions in velocity and correspondingly large increases
in force, *worm* gearing is used (Fig. 72*c*).   The worm is cylindrical in
shape with a continuous tooth form describing a helix on the lateral
surface of the cylinder.   Action takes place by the worm driving the
gear.   The teeth of the gear are set at the helical angle and are usually
made concave to fit into the convex surfaces of the worm.

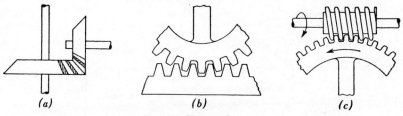

(a)   *(b)*   *(c)*

FIG. 72. Types of gears.

68. Materials and Design Stresses.   Among the materials used
for gears are cast iron, steel (forged or cast), fabroid (woven cotton cloth
used as a base), rawhide, and Bakelite.   While many gears are made
of cast iron, steel is preferred and is becoming more generally used as a
tooth material for heavy duty because of the inability of cast iron to
withstand shock.   However, the wearing quality of cast-iron teeth is
superior.   Gears made of alloy steels are often chromium-vanadium
alloys.   Alloy bronzes, mostly nickel bronze and to a lesser extent
phosphor bronze, are being used increasingly.

Cast-iron and steel gears are often hardened by heat-treatment to
increase the resistance to wear.   Chilled castings are applicable to
cast-iron gears, casehardening is used for low-carbon steel, and heating
and quenching (with tempering) for medium grades.   Because of the
greater or less degree of shock which all gear teeth must resist, it is
important that the toughness of the material be retained in the heat-
treatment.

In determining the design-stress value for the gear teeth, factors of
shock and repeated stress should be considered.   Owing to the fact

that gear teeth are not perfectly formed, the contacts between the teeth of two mating gears are not the same at all points. This condition leads to a slight degree of impact and attending shock. Rapid starting and stopping of the mechanism also contribute to shock. Second, each gear tooth is stressed for only a small fraction of the time of each revolution. Hence, the teeth are subjected to repeated load applications and repeated stresses are induced. To guard against fatigue failure, it is advisable to use the endurance limit value for the limit stress in design-stress calculations.

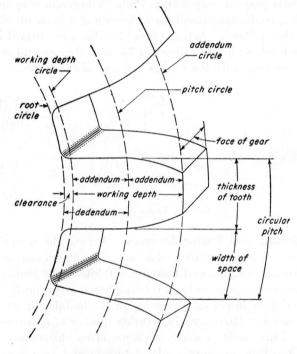

Fig. 73. Parts of a gear tooth.

69. Parts of a Gear Tooth. The names of the various parts of a spur gear are shown in Fig. 73. The *pitch circle* is the foundation for all calculations. Pitch circles of two mating gears are tangent to one another, as in the case of the circles of Fig. 71. The distance from the pitch circle to the tip of the tooth is the *addendum*. Two addendum distances are added together to form the *working depth*. The purpose of the *clearance* is to prevent contact at the base of the tooth with the top of the mating gear tooth. Addendum and clearance distance added is the *dedendum*.

A gear is known by its *pitch*. The term *pitch* has two meanings, namely, *diametral pitch* and *circular pitch*. Diametral pitch P, the more important pitch for purposes of gear identification, equals the number of teeth N divided by the pitch diameter D, or, as sometimes expressed, the number of teeth for every inch of pitch diameter.

$$P = \frac{N}{D} \tag{40}$$

When the unqualified term *pitch* is used, diametral pitch is meant. Circular pitch p is the distance, measured along the pitch circle, from a point on a tooth to the corresponding point on the next tooth. This distance multiplied by the number of teeth in the gear equals the circumference of the pitch circle. That is,

$$p \times N = \pi D$$

and, by transposition,

$$p = \frac{\pi D}{N} \tag{41}$$

By definition it follows that the circular pitch of all mating gears must be equal. Likewise it is also true that the diametral pitch for all such gears is equal, because of the fact that the numerator and denominator of the fraction N/D change proportionately, thus keeping the value of the fraction the same.

Increasing the number of teeth with a constant pitch diameter means a decrease in the tooth size and, from Eq. (40), we see that this also means an *increase* in the value of the diametral pitch. In contrast, we see, from Eq. (41), that, for a constant diameter, increasing the number of teeth (decreasing the tooth size) means a *decrease* in the circular pitch. It is, therefore, well to remember that diametral pitch varies *inversely* with the tooth size and circular pitch varies *directly* with the tooth size. Figure 74 shows size of gear teeth with corresponding values of diametral pitch.

Now let us multiply the left-hand side of Eq. (40) by the left-hand side of Eq. (41) and equate the product to the product of the right-hand sides of these same equations.

$$P \times p = \frac{N \times \pi \times D}{D \times N}$$

which by cancellation reduces to

$$Pp = \pi \tag{42}$$

The addendum a is easily found if the diametral pitch is known, because one is always the reciprocal of the other.

$$a = \frac{1}{P} \tag{43}$$

The clearance c also can readily be determined from the diametral pitch, since

$$c = \frac{1}{8} \times \frac{1}{P} = \frac{1}{8P} \tag{44}$$

Students should become thoroughly familiar with the foregoing definitions and their interrelationships.

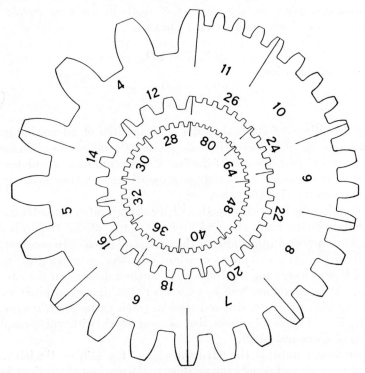

Fig. 74. Tooth size and diametral pitch.

When the letters G or P are used as subscripts to a symbol, the term for which that particular symbol stands is for the gear or the pinion, as for example, D_G denotes the pitch diameter of the gear and N_P denotes the number of teeth in the pinion.

Illustrative Example. Two spur gears are to be used to increase the rpm as nearly as possible $3\frac{3}{4}$ times. The diametral pitch is 7 and the shaft centers should be as close as possible to $10\frac{1}{2}$ in. Calculate the exact velocity ratio and the exact distance center-to-center of gears.

To increase the rpm, the driver is the gear and the driven is the pinion.

Let R_G equal the pitch radius of gear and R_P equal the pitch radius of pinion.

$$\frac{R_G}{R_P} = 3.75$$

$$R_G = 3.75R_P$$

also

$$R_G + R_P = 10.5$$

Then, by substitution,

$$3.75R_P + R_P = 10.5$$
$$4.75R_P = 10.5$$

and

$$R_P = \frac{10.5}{4.75} = 2.21$$

also

$$R_G = 3.75 \times 2.21 = 8.29$$

To check,

$$8.29 + 2.21 = 10.50$$
$$N_P = PD_P = 7 \times 4.42 = 30.94 \qquad (40)$$

Make 31 teeth for pinion.

$$31 \times 3.75 = 116.25$$

Make 116 teeth for gear.

$$\text{Actual velocity ratio} = {}^{11}\!\%_1 = 3.74, \text{ or } 1{:}3.74$$

The actual pitch diameters are

$$D_G = \frac{N_G}{P} = \frac{116}{7} = 16.57 \text{ in.}$$

$$D_P = \frac{N_P}{P} = \frac{31}{7} = 4.43 \text{ in.} \qquad (40)$$

The actual distance center-to-center of gears is

$$\frac{16.57 + 4.43}{2} = \frac{21.00}{2} = 10.50 \text{ in. (approx)}$$

70. The Involute System of Gearing. The outline of the outer portion of the gear tooth follows a special curve. The curve in most general use is called an *involute*. Usually the $14\frac{1}{2}°$ involute system is used, although there are definite advantages to the use of the 20° involute. Figure 75a shows an involute curve. Imagine that a string is wound around the circumference of a circle and is tied to a pencil at the end. If the pencil is held with point on paper and the string is gradually unwound, but always kept taut, the line made by the pencil is an involute curve.

To construct a $14\frac{1}{2}°$ involute curve for a gear tooth, draw a tangent to the pitch circle at any point as A (Fig. 75b). Through this point

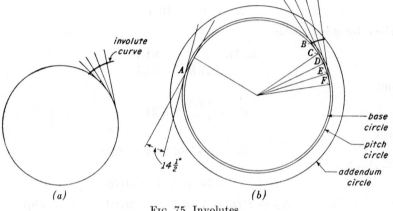

FIG. 75. Involutes.

draw a line to make an angle of $14\frac{1}{2}°$ with the tangent. Drop a perpendicular from the center of the pitch circle to the line just drawn. This perpendicular now becomes the radius of the *base circle* from which the involute is made. Select some other point on the base circle as B, and lay off short equal arcs BC, CD, DE, and EF. Draw radii from points C, D, E, and F and tangents perpendicular to the radii. The second point on the involute (point B is the first) is found by laying off the length of the arc CB on the tangent from point C. The next point is found by laying off the length of the arc DB on the tangent from point D. Other points are obtained similarly.

Drawing involute curves from circles with ever-increasing diameters results in flatter and flatter involutes. A circumference with an infinite diameter becomes a straight line, and the involute from such a circumference would also be a straight line. Hence, the tooth sides of a rack are straight lines.

The portion of the tooth outline between the base circle and the working-depth circle does not follow any definite curve. For a set of mating gears (all of which have the same diametral pitch), those with the larger diameters (larger number of teeth) have teeth slightly thicker at the working-depth circle than at the base circle, while the reverse is true for those gears with smaller diameters (smaller number of teeth). This variation is designed to avoid tooth interference.

71. Design of a Spur-gear Tooth. We are now ready to study the design proportions of a spur-gear tooth. It is assumed that the tooth acts as a cantilever beam and the tooth thrust P_1 (Sec. 47) is the force acting at the end of the tooth beam along the entire width of the tooth, as illustrated in Fig. 76. Furthermore, the direction of P_1 is considered in this analysis to be perpendicular to the longitudinal axis of the beam. Because the fillets add greatly to the strength of the tooth at the base, the beam is regarded as starting at the working-depth circle. Hence, the span is the working depth equal to twice the addendum. In a cantilever beam the dangerous section is at the fixed end which in our case is the section at the working-depth circle. One further assumption is that the dimensions of the tooth at this dangerous section are the same as at the pitch circle, namely, the thickness of tooth as the depth dimension and the face of gear as

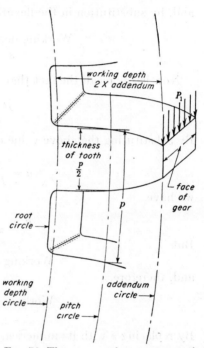

FIG. 76. The gear tooth acts as a cantilever beam.

the width dimension of the beam. Since in new gears the thickness of tooth and width of space (Fig. 73) are almost equal, each of these dimensions can be represented by $p/2$ (again Fig. 76).

The flexure formula for beams, $M/s = I/c$, can then be applied. For a rectangle, $I = bh^3/12$ (table on page 215), where b is the dimension parallel to the neutral axis (width dimension) and h is the dimension perpendicular to that axis (depth dimension). Also since $c = h/2$, then

$$\frac{I}{c} = \frac{bh^3}{12} \times \frac{2}{h} = \frac{bh^2}{6}$$

and the flexure formula for rectangular beams can be written

$$\frac{M}{s} = \frac{bh^2}{6} \tag{45}$$

which form is somewhat more convenient to use.

The maximum bending moment for the tooth beam is

$$\text{Working depth} \times P_1$$

and, by substitution in the flexure formula,

$$\frac{\text{Working depth} \times P_1}{s} = \frac{bh^2}{6}$$

Now let us recall the fact that

$$P = \frac{\pi}{p} \tag{42}$$

By substituting the above value of P in

$$a = \frac{1}{P} \tag{43}$$

we have

$$a = \frac{1}{\pi/p} = \frac{p}{\pi}$$

But

$$\text{Working depth} = 2a$$

and, therefore,

$$\text{Working depth} = \frac{2p}{\pi}$$

By replacing π with its numerical value, the equation reduces to

$$\text{Working depth} = 0.64p$$

This value is then substituted in the flexure formula

$$0.64pP_1 = \frac{sbh^2}{6}$$

The face of the gear (b dimension) usually varies between $1\frac{1}{4}$ to $3\frac{1}{2}$ times the circular pitch. For the present let us say that $b = 1\frac{1}{4}p$, or $5p/4$. Also since $h = p/2$, we have, by substitution,

$$0.64pP_1 = s \times \frac{5p}{4} \times \left(\frac{p}{2}\right)^2 \times \frac{1}{6}$$

which, by combination and the cancellation of p from each side of the equation, reduces to

$$0.64P_1 = \frac{5sp^2}{96}$$

Then, by solving for p,

$$p = \sqrt{\frac{(0.64P_1)(96)}{5s}}$$

and

$$p = 3.51 \sqrt{\frac{P_1}{s}}$$

This equation is applicable only when the value of the face of gear is $1\frac{1}{4}$ times the circular pitch. For varying relationships the number 3.51 will vary. If we substitute the symbol k for this number, we have a general formula applicable to these varying relationships, namely,

$$p = k \sqrt{\frac{P_1}{s}} \tag{46}$$

where values of k are given in Table 6.

TABLE 6

Face of Gear (width of tooth)	Values of k
$1\frac{1}{4}p$	3.51
$1\frac{1}{2}p$	3.20
$1\frac{3}{4}p$	2.97
$2p$	2.78
$2\frac{1}{4}p$	2.62
$2\frac{1}{2}p$	2.48
$2\frac{3}{4}p$	2.37
$3p$	2.27
$3\frac{1}{4}p$	2.18
$3\frac{1}{2}p$	2.10

After the circular pitch p has been calculated from Eq. (46), the other important gear dimensions can be calculated by the use of Eqs. (40) through (44). The design proportions of spur gears usually required in problems of machine design are:

1. Pitch diameter
2. Diametral pitch
3. Number of teeth
4. Outside (over-all) diameter
5. Face of gear

The diametral pitch as determined by calculation will probably not come out exactly to a value for which a standard form cutter is available. It should then be changed to the next standard *lower* value. Why is not the next higher value taken? Table 7 shows these standard values. Refer to Fig. 74 for the tooth sizes.

TABLE 7

Variation	Diametral pitch
¼	1, 1¼, 1½, 1¾, 2, 2¼, 2½, 2¾, 3, 3¼, 3½, 3¾
½	4, 4½, 5, 5½
1	6, 7, 8, 9, 10, 11, 12, 13, 14, 15
2	16, 18, 20, 22, 24, 26, 28, 30
Misc.	32, 36, 38, 40, 44, 48, 50, 56 60, 64, 70, 80, 120

Illustrative Example. Shaft A is to transmit 10 hp to shaft B by means of mating nickel-bronze spur gears. The driver, A, will rotate at 150 rpm and the driven, B, at 50 rpm. The distance between center to center of the shafts is 16 in. (or as nearly as possible). Calculate the principal design proportions of the gear teeth. Assume the face of gear to be three times the circular pitch and the design stress to be 4500 psi.

To reduce the speed, the pinion is the driver and is fitted on shaft A. Also the gear diameters must be in the ratio of 1:3.

$$R_G + R_P = \text{center distance}$$
$$\frac{R_G}{R_P} = \frac{150}{50} = 3$$
$$R_G = 3R_P$$

By substitution,
$$3R_P + R_P = 16$$
$$4R_P = 16$$
$$R_P = 4.0 \text{ in.}$$
$$R_G = 3 \times 4 = 12.0 \text{ in.}$$

Hence
$$D_P = 8.0 \text{ in. (tentatively)}$$
$$D_G = 24.0 \text{ in. (tentatively)}$$
$$T = \frac{63,000H}{n} = \frac{63,000 \times 10}{150} = 4200 \text{ lb-in.} \quad (11)$$

$$P_1 = \frac{T}{R_P} = \frac{4200}{4} = 1050 \text{ lb}$$

$$p = k\sqrt{\frac{P_1}{s}} = 2.27 \times 0.484 = 1.10 \text{ in.} \qquad (46)$$

$$P = \frac{\pi}{p} = \frac{3.14}{1.10} = 2.85 \qquad (42)$$

Make $P = 2.75$ (Table 7).

$$N_P = PD_P = 2.75 \times 8 = 22 \qquad (40)$$

$$a = \frac{1}{P} = \frac{1}{2.75} = 0.36 \text{ in.} \qquad (43)$$

The new value of the circular pitch is

$$p = \frac{\pi}{P} = \frac{3.14}{2.75} = 1.14 \text{ in.} \qquad (42)$$

Outside diam of pinion = 0.36 × 2 + 8.00 = 8.72 in.
Outside diam of gear = 0.36 × 2 + 24.00 = 24.72 in.
Face of gear = 3 × p = 3 × 1.14 = 3.42 in.

PRINCIPAL DESIGN PROPORTIONS OF THE GEAR TEETH

	Pinion	Gear
Pitch diam................	8.00 in.	24.00 in.
Diam pitch................	2.75	2.75
Number of teeth...........	22	66
Outside diam..............	8.72 in.	24.72 in.
Face of gear..............	3.42 in.	3.42 in.

72. The Lewis Equation. The so-called *Lewis equation* embodies a variation of the method of proportioning discussed in Sec. 71. The tooth is again considered to be a cantilever beam with the dangerous section at the base of the tooth (Fig. 77), and the development as before based on the flexure formula for rectangular beams, namely,

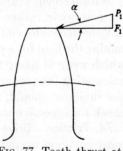

$$M = \frac{sbh^2}{6} \qquad (45)$$

In the previous analysis, we assumed that the tooth thrust P_1 acted perpendicularly to the longitudinal axis of the tooth beam. Strictly

Fig. 77. Tooth thrust at an angle.

speaking, this is not true, and P_1 acts at an angle α (Fig. 77) to the perpendicular. The angle α, called the *pressure angle*, is the same as the involute system used. For example, for a $14\frac{1}{2}°$ system, α is $14\frac{1}{2}°$. Lewis considers the perpendicular component F_1 of P_1 as the force causing bending in the tooth. The Lewis equation states that

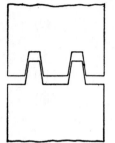

$$F_1 = sbpy \qquad (47)$$

where y is a variable called the *Lewis factor* and the other terms are as previously designated. The value of y depends on the number of teeth in the gear, the height of tooth, and the pressure angle. Tables are available containing values of y for differing conditions.

In the Lewis equation, both p and y depend on the number of teeth. Therefore, in design we are confronted with two unknowns and the equation cannot be solved directly. A trial value for y is chosen from the table and p calculated. If the choice was reasonable, no further calculations are needed.

FIG. 78. One type of friction drive.

73. Wear of Gear Teeth. When requirements call for day in and day out service of a machine, the wear of the gear teeth presents a problem equal to that of their strength. Under heavy service conditions the teeth of mating gears may become so worn that they are no longer serviceable. Among the causes of wear is the presence of grit and minute particles which have been broken off the teeth, as well as improper or insufficient lubrication. Mention already has been made of various forms of heat-treatment for hardening the surfaces to minimize wear.

Within recent years, design formulas based on gear wear rather than on gear strength have been developed for the purpose of determining the load for a particular tooth beyond which wear is likely to be excessive. After calculating the proper size tooth for strength, the designer should, if conditions warrant, check for excessive wear.

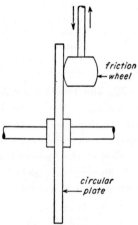

FIG. 79. Friction drive with variable velocity ratio.

74. Friction Drives. Undoubtedly the most common example of a friction drive is the driving wheels of an automobile. These wheels are able to impart motion to the vehicle because of the large amount of friction between tire and road. In wet and icy weather, when the coefficient of friction has been very considerably reduced, it is another story.

Applications of friction drives are usually limited to small torques. Covering the contact surfaces of the friction wheels with rubber gives both a high friction coefficient and a fairly durable wearing surface. The frictional resistance is also increased by shaping the surfaces to fit into one another, as shown in Fig. 78.

In Fig. 79 we have a friction-drive mechanism the speed of which can be changed without interrupting the transmission of power. The *linear* speed of the circular plate increases with the distance from its center. Therefore, as the friction wheel is moved upward or downward by a pinion-and-rack device, the rpm of the wheel will increase or decrease.

PROBLEMS

1. Calculate the outside diameter, root diameter, number of teeth, and circular pitch for a gear with a pitch diameter of 22 in. and a diametral pitch of $5\frac{1}{2}$.

2. A certain gear has an outside diameter of $6\frac{5}{8}$ in. (as closely as can be measured) and has 91 teeth. Determine its probable diametral pitch.

3. A gear of 6 diametral pitch meshes with a pinion. Center distance of gears is 9 in. Speed ratio is $3\frac{1}{2}:1$. Calculate:

a. The pitch diameter of the pinion
b. The number of teeth in the pinion and the gear

4. Calculate the outside diameter, pitch diameter, and number of teeth on each of two mating spur gears. Use the following specifications. The center-line distance between shafts is to be approximately $6\frac{3}{4}$ in., the diametral pitch 10, the speed of the pinion 1750 and that of the gear 500 rpm.

5. Two gears are to mesh together. The distance between their centers is 4 in. One gear is to make 11 revolutions while the other makes 5. Diametral pitch is 10. Calculate the following:

a. Pitch diameter of each gear
b. Outside diameter of each gear
c. Height of tooth
d. Circular pitch

6. Calculate the distance between centers of two mating spur gears with a velocity ratio of $1:2.3$, and a diametral pitch of $3\frac{1}{2}$. The number of teeth in the pinion is 30.

7. Two mating spur gears have 108 and 24 teeth and were cut with a standard cutter. Rough measurement gives a distance of $\frac{1}{2}$ in. from center to center of teeth.

a. What is the probable correct diametral pitch?
b. What is probably the correct circular pitch?
c. What should be the distance between centers of gears?

8. A gear and pinion are designed for an exact increase in velocity in the ratio of $1:4$. The tooth design calls for a diametral pitch of 4. The centers of the gears should be as close as possible to 13 in. What is this exact distance to be?

9. Two mating spur gears should have a velocity ratio of $1:5\frac{1}{4}$ (as close as possible but not less) and a distance center to center of 16 in. (as close as possible). The diametral pitch is $3\frac{1}{2}$. Calculate the exact velocity ratio and distance center to center of gears.

10. A worm and worm gear have a velocity ratio of 55:1. The pitch diameter of the worm is 3 in. and the worm tooth advances 0.393 in. for one revolution. What must be the pitch diameter of the worm gear?

11. Compare a $14\frac{1}{2}°$ involute gear tooth with a 20° tooth in regard to

a. The position of the base circle

b. The appearance of the involute curve which forms the sides of each tooth

12. Make calculations to determine whether a gear tooth with a diametral pitch of 3 on an 11-in.-diam spur gear is sufficiently strong to transmit 20 hp at 600 rpm. The face of gear is three times the thickness of tooth. Use a design stress of 3000 psi.

13. An 8-in.-diam steel spur gear has teeth of 7 diametral pitch and a width of face equal to twice the circular pitch. Calculate the speed at which this gear must rotate to transmit safely 5.5 hp. Use a design stress of 6500 psi.

14. A 12-in.-diam steel spur gear transmits a torque of 3200 lb-in. Calculate the principal design proportions of the gear tooth as given in this chapter. Assume that the width of face equals $2\frac{1}{2}$ times the thickness of the tooth, and use a design stress of 8000 psi.

15. Two shafts running at 100 and 200 rpm transmit 75 hp. They are connected by spur gears as near as possible to 60 in. between centers. Each gear has six arms.

a. Calculate the proper size tooth to resist the thrust, assuming the face width to be three times the circular pitch and a design stress of 4500 psi.

b. If an elliptical section is to be used in the gear arms, determine the maximum depth and width of section of the pinion arm. Assume one-half of the number of arms to be acting to resist the load, and major and minor axes in the ratio of 2:1.

16. Refer to the accompanying figure.

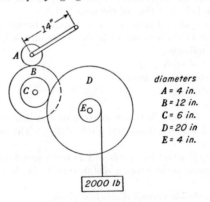

diameters
A = 4 in.
B = 12 in.
C = 6 in.
D = 20 in
E = 4 in.

2000 lb

a. Calculate the necessary tooth size for the teeth on gear D. Assume the face width as $3\frac{1}{2}$ times the circular pitch.

b. With an efficiency of 72 per cent, what force is necessary on the hand crank to raise the weight?

17. A steel gear and pinion transmit 50 hp and reduce the velocity in the ratio of 4.6:1. The speed of the driver is 900 rpm and distance between shaft centers is 28 in. Calculate the principal design proportions for the teeth of both gear and pinion. Use a design stress of 7000 psi, and assume that the ratio of tooth-beam depth to width is 1:5.

18. Calculate the proper diametral pitch of a gear of 22 in. pitch diam to transmit a torque of 5400 lb-in. Use the Lewis formula. Assume the design stress as 4000 psi, the angle α as $14\frac{1}{2}°$, the width as $2\frac{3}{4}$ times the circular pitch, and the Lewis factor (y) as 0.115. Recalculate the diametral pitch by the method developed in the text and compare results.

19. The circular plate of Fig. 79 is revolving at 100 rpm. The diameter of the friction wheel in contact is $\frac{3}{4}$ in. maximum. How far up must it be moved to increase its velocity by 200 rpm?

CHAPTER 14

COUPLINGS

75. Couplings for Collinear Shafts. A coupling is used to connect the ends of two shafts in order that they can act as one continuous member. Although most frequently the axes of the shafts are collinear (would coincide if extended), couplings have also been devised for shafts in which the axes are noncollinear, sometimes parallel, sometimes not. We shall confine our discussion mainly to couplings for collinear shafts.

Figure 80a and b shows early types of couplings, the muff coupling and the flange coupling, respectively. The muff coupling, relying

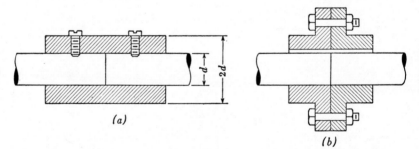

(a)

(b)

Fig. 80. Types of coupling. (a) Muff coupling; (b) flange coupling.

as it did upon the shearing strength of the setscrews exclusively, was not well adapted for heavy duty. The two parts of the flange coupling are held together by bolts in a circle through the flanges. A key fixes the coupling to the shaft. This coupling is more satisfactory in many respects, but, in its early form, owing to the protruding nuts and boltheads on the revolving flange, was found to be a cause of accidents, and was subsequently condemned by the New York State Workmen's Compensation Commission. This condition led to the development of the safety flange coupling in which the flange is made

with a lip around the circumference, protruding at right angles to the
face, as illustrated in Fig. 81a and b. A piece of sheet metal in the
shape of a ring can be inserted on each side between the hub and the
lip, as shown in Fig. 81b. In this way the revolving surfaces are
made smooth throughout and the dangerous feature of the flange
coupling completely eliminated. Flange couplings are usually made
of cast iron.

76. Design of the Safety Flange Coupling. We will recall that any
cross section of a shaft in resisting a torque develops a stress of torsion
which is in reality a stress of shear. To one side of a given section, a
twisting moment is driving the shaft; to the other side, for equilibrium,
an equal and opposite moment is resisting the driving moment. The

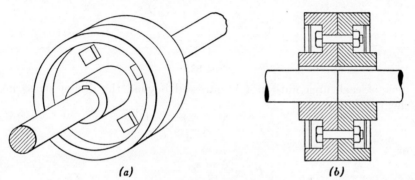

(a)   (b)

FIG. 81. Safety flange coupling.

cross-sectional area must resist the shearing tendency. In like manner, the bolts in the coupling must resist this shearing tendency
because the cross-sectional area of the bolts takes the place of the
cross-sectional area of the shaft at the joint.

Bolts with diameters $\frac{1}{16}$ in. less than the diameters of the holes
are used to allow for slight variations in the positions of the matching
holes. These inaccuracies cause the flange of a flange coupling, when
in use, to bear on some of the bolts only, resulting in these bolts
resisting the shearing forces while the others do not. For this reason,
it is better practice to consider the number of bolts acting to be one-
half of the total number of bolts in the coupling. As in the arm pulley,
when there are three or four bolts, two are considered to be acting, and
when there are five or six bolts, three are so considered.[1]

---

[1] However, when the holes in each part of the flange are reamed to match the
holes exactly, as well as to allow a close fit of the bolts, all the bolts may be considered to be resisting the shearing forces.

The value of the shearing force P_1 at the bolt circle of radius r (Fig. 82) is the torque T on the shaft divided by r.

$$P_1 = \frac{T}{r}$$

Each bolt considered as acting is resisting with its cross-sectional area $\pi d^2/4$, where d is the diameter of the bolt. For safety, each square

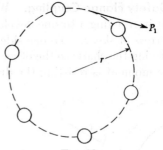

FIG. 82.

inch of each area must not be stressed beyond the design stress in shear[1] s_s. Hence,

$$P_1 = \frac{N}{2} \times \frac{\pi d^2}{4} \times s_s$$

or

$$P_1 = \frac{N \pi d^2 s_s}{8}$$

where N is the total number of bolts. By transposition,

$$N = \frac{8P_1}{\pi d^2 s_s} \tag{48}$$

In addition to the shearing stress in the bolts, the flange is stressed in bearing as it in turn resists the push of the bolts. This is similar to the stress of bearing in a hub when pushed by a key (Sec. 57). The resisting area in a plate to the push of a bolt or a rivet is taken as the diameter of the bolt or rivet multiplied by the thickness of the plate. If t represents the thickness of the flange, the area resisting the bearing stress for each bolt is $t \times d$. Again, for safety, each square inch of this area must not be stressed beyond the design stress of the steel in bearing s_b. Therefore,

$$P_1 = \frac{N}{2} \times td \times s_b$$

---

[1] Although the bolts are of steel, a very low design stress should be assigned for the reasons outlined in Sec. 99.

and

$$P_1 = \frac{Ntds_b}{2}$$

also, by transposition,

$$t = \frac{2P_1}{Nds_b} \qquad (49)$$

There is also a possibility, though not a likelihood, of the flange failing in shear similar to the type of failure investigated in the solid-web pulley (Sec. 62). Such failure would be on a circular surface next to the hub. The required flange thickness as calculated for either bearing or shear is usually very small. When checking a standard coupling as selected from a manufacturer's catalogue for strength, calculations for both bearing and shearing in flange may be omitted with safety.

Transverse loads inducing bending stress in the shaft will also induce bending stress in the flange of the coupling. As in the shaft, this stress reverses itself during rotation, since any point in tension at one position will be in compression when rotated 180°. Again as in shafts, if such stress exceeds the endurance limit of the material, fatigue failure will eventually result. However, the flange of a standard coupling is of such thickness that, except in very unusual cases, safety from fatigue failure is ensured.

Further considerations in the design of a safety flange coupling are:

1. The hub diameter by rule of thumb varies from 2 to $2\frac{1}{2}$ times the diameter of the shaft.

2. The key is designed according to Sec. 57. Since *each* part of the coupling transmits the entire torque, there must be sufficient length of key in each part to resist this torque. The length of key governs the length of hub. Each part has a separate key.

3. There must be sufficient clearance between the lip and the hub to allow for the use of a socket wrench in tightening the bolts. It is usual to make the lip the same thickness as the flange.

After making a selection from a manufacturer's catalogue, it is good practice to make computations to check the coupling selected for strength.

Illustrative Example. A 2-in.-diam solid steel shaft transmits a torque of 9400 lb-in. The shaft is to be lengthened by means of a cast-iron safety flange coupling. A tentative selection has a flange thickness of $\frac{3}{4}$ in., a 4-in.-diam hub, a 6-in.-diam bolt circle, and has holes for six $\frac{3}{4}$-in. bolts. Each hub is 4 in. long and is cut for a $\frac{1}{2}$-in. square key. Make calculations to determine:

   a. The number of $\frac{3}{4}$ in. bolts needed
   b. The required thickness of the flange
   c. The required length of the hub

Use design-stress values of 3000 psi in shear for the bolts, 5000 psi in shear and 18,000 psi in bearing for the cast-iron flanges, and 10,000 psi in shear for the steel key.

   a.

$$P_1 = \frac{T}{r} = \frac{9400}{3} = 3130 \text{ lb}$$

$$N = \frac{8P_1}{\pi d^2 s_s} = \frac{8 \times 3130}{3.14 \times 0.75 \times 0.75 \times 3000} = 4+ \quad (48)$$

Use six bolts for symmetry.
   b. For bearing in flanges,

$$t = \frac{2P_1}{N d s_b} = \frac{2 \times 3130}{6 \times 0.75 \times 18,000} = 0.0772 \text{ in.} \quad (49)$$

For shear in flange,

$$P_1 = \frac{T}{r} = \frac{9400}{2} = 4700 \text{ lb}$$

$$t = \frac{P_1}{\pi d s} = \frac{4700}{3.14 \times 4 \times 5000} = 0.0748 \text{ in.} \quad (33)$$

   c.

$$P_1 = \frac{T}{r} = \frac{9400}{1} = 9400$$

$$l = \frac{2P_1}{t s_b} = \frac{2 \times 9400}{0.5 \times 18,000} = 2.09 \text{ in.} \quad (32)$$

Since in the key investigation the design stress for bearing is less than double the design stress for shear, the greatest key length results from the bearing calculation (Sec. 57).

The flange coupling tentatively chosen is sufficiently strong in every respect.

77. Couplings for Noncollinear Shafts. Although parallel, two shafts may for one reason or other be slightly offset. The *Oldham coupling* (Fig. 83*a*) can be used to join such shafts. As in the flange coupling, two flanges are keyed to the ends of the shafts. A third part, in the shape of a disk, fits between the flange faces. Two grooves, one on each face of the disk and mutually perpendicular, act as key seats. Each flange has a key to fit into the key seats loosely. Figure 83*a* shows the three parts disassembled. Because of the misalign-

ment of the shafts, the keys are continually sliding back and forth
in the seats as the shafts rotate. To keep the mechanism engaged,
axial movement of both shafts must be prevented.

When the axes of the shafts are not parallel but do meet at a point,
the shafts can be connected by a *universal joint* or *coupling*, shown disassembled in Fig. 83*b*. The two clevises are fastened together by steel

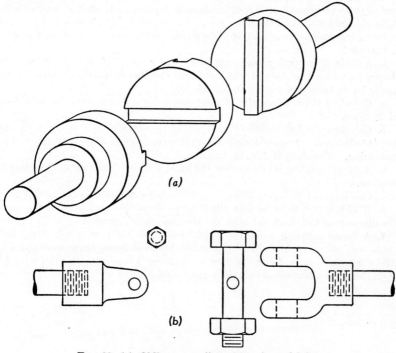

(a)

(b)

FIG. 83. (*a*) Oldham coupling; (*b*) universal joint.

pins, **or** bolts, the smaller pin passing perpendicularly through a hole
in the larger one. When rotating, the point marking the intersection
of the axes is thus maintained in one position and bending is avoided.
Universal joints are used where the eccentricity is as much as 20° or
even more, but are not suitable for heavy torques.

PROBLEMS

1. What is the maximum horsepower that can be transmitted safely by a flange
coupling with six ⅜-in.-diam bolts in a 3-in.-diam bolt circle, when revolving at
1160 rpm? Use a design stress of 3000 psi for the bolts.

2. Calculate the number of ⅝ in. bolts needed in a cast-iron flange coupling
to develop sufficient shearing strength to resist safely a net belt pull of 640 lb at

the circumference of a 34-in.-diam pulley. Assume a design stress in shear for the bolts as 4000 psi and that all the bolts are 3 in. away from the shaft center.

3. A shaft transmits 125 hp at a speed of 270 rpm. The ends are connected by a flange coupling with a flange thickness of $\frac{1}{2}$ in. and with $\frac{3}{4}$-in. bolts in an 8-in. bolt circle. Make calculations to determine the number of bolts needed and to check the flange thickness. Use a design stress of 3500 psi for the bolts in shear and 14,000 psi for the flange in bearing.

4. A flange coupling is used to connect two sections of a line shaft. Keyed to one of these sections there is a 42-in.-diam pulley with a tension of 500 lb on the tight side and 200 lb on the slack side. The coupling is to have six $\frac{1}{2}$-in.-diam bolts with a design stress in shear of 3000 psi. What must be the diameter of the bolt circle?

5. A 66-in.-diam pulley transmits 220 hp at a speed of 500 rpm to a 22-in. driven pulley on the shaft of which a flange coupling is to be attached. There are to be six bolts in a 6-in.-diam bolt circle.

a. Calculate the required size of the bolts, using a design stress for bolt shear of 2800 psi.

b. Calculate the hub length of each part of the coupling, if a $\frac{1}{2}$- by $\frac{3}{8}$-in. flat key is to be used. Assume design-stress values of 8000 psi in shear and 20,000 psi in bearing. The shaft is 2 in. in diameter.

6. A flange coupling is to transmit 180 hp at 120 rpm from one shaft section to another.

a. Calculate the necessary shaft diameter, using a design stress of 8500 psi.

b. If 12 bolts are to be used on a bolt circle, the diameter of which is $3\frac{1}{2}$ times the diameter of the shaft, calculate the necessary bolt diameter.

7. A flange coupling made for eight $\frac{3}{8}$-in.-diam bolts on a 4-in.-diam bolt circle was selected for a shaft connection, but was found to be unobtainable. Make calculations to determine whether a flange coupling with four holes for $\frac{1}{2}$-in. bolts on a 6-in.-diam bolt circle can be substituted with safety.

CHAPTER 15

BEARINGS

78. Bearings and Journals. A *bearing* is a general term used to denote a support for a moving machine part. Most of the time the part supported describes a rotary motion as in the case of a rotating shaft. Sometimes, as in a typewriter carriage, the motion·is along a line (rectilinear) and the support is also termed a *slide*, or *way*.

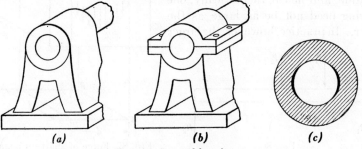

(a)   *(b)*   *(c)*

FIG. 84. Journal bearings.

Figure 84*a* and *b* show simple one-piece and two-piece *journal* bearings for rotating shafts. The part that holds the bearing to the base of the machine is called a *bracket*. In Fig. 84*c* there are shallow grooves at the mid-section of the bearing to retain a supply of oil. Grooves are cut in various ways for this purpose. The portion of the shaft within the bearing is the *journal*. Some bearings encase the journal completely, as do those of

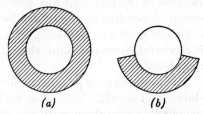

(a)   *(b)*

FIG. 85. Full and partial bearings.

Fig. 84, whereas others do so only partially. These are known as *full bearings* and *partial bearings*, respectively, and are shown in Fig. 85*a* and *b*. To allow for assembly clearance and lubrication, the bearing diameter must be somewhat larger than that of the shaft itself.

Bearings can also be classified as sliding contact or rolling contact bearings. *Roller* and *ball* bearings are of the latter type. Roller bearings are made up of sets of cylinders (or sometimes cones) placed between stationary and moving parts. In ball bearings, the rolling parts are spheres.

79. Design of a Sliding-contact Bearing. As previously discussed, the forces of a belt on a pulley, the thrust of a gear tooth on a mating gear, the push of a follower on a cam, as well as the weight of the part itself, cause the shaft to act as a loaded beam. Therefore, the bearing is subjected to a force equal to the beam reaction and should be designed to support this force. In ordinary cases, the force of the reaction is applied perpendicularly to the axis of the shaft. Figure 86 represents the loading diagram of a typical shaft beam. R_1 and R_2 are unequal in value, and hence, theoretically, one bearing need not be as large as the other. In practice, however, the larger

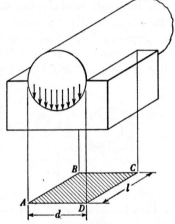

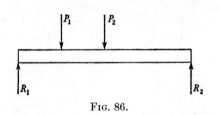

Fig. 86.   Fig. 87. Force of shaft on bearing.

reaction is used as the basis for the design of both bearings. In this way design work is decreased and the parts can be interchanged. Because of the fact that any loaded beam deflects to a greater or less extent, depending on the loads, the push of the journal on the bearing is greater nearer to the loads. In the design of the bearing, this fact is rarely taken into account, that is, the shaft is regarded as a perfectly rigid body.

A journal bears or pushes on the curved surface of the bearing, as is shown in Fig. 87. Maximum push is at the extreme bottom of the semicircle, where the direction of the force is represented by the arrow perpendicular to the surface of the bearing. To the right and left of this point, there is less push and the tendency for the shaft to slip past the bearing increases. At the top of the bearing shown, the shaft does not push on the bearing at all. Because of this variation, the projection $ABCD$ of the curved surface is considered as the equivalent resisting area. Furthermore, the stress is considered as equally distributed on this projected area. This theory is similar to that used in the bearing investigation of the flanged coupling (Sec. 76). It follows that the surface must be of such size that each square inch is not stressed excessively when resisting the reaction force R. In symbols,

$$R = dls$$

where d = diameter of shaft
l = length of bearing (journal)
s = special design stress for shaft bearings (allowable bearing pressure)

The shaft diameter has previously been determined and, as the special design stress is assumed, the equation can be solved for l, the only unknown term,

$$l = \frac{R}{sd} \tag{50}$$

What value is to be taken for this special design stress? To answer this question, let us briefly consider the lubricating action of the oil supplied to the bearing. Even though the shaft is pushing down on the bearing, still a thin film of oil normally entirely surrounds the shaft while in motion (Fig. 88). This condition is made possible by the natural adhesion between the oil and the metals of the parts. Without such an oil film, excessive wear would very shortly take place. Too much pressure of the shaft on the bearing squeezes out the oil with the result just indicated. An added factor in limiting the pressure to be placed on the bearing is the heat generated by the rotation of the shaft, the amount of which depends both

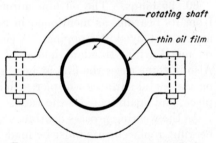

Fig. 88. Oil film around rotating shaft.

on the pressure and the speed of rotation. If the bearing is of insufficient size to carry away this generated heat, the increased temperature will reduce the viscosity (stiffness) of the oil and again increase the tendency of the shaft to squeeze the oil out of the bearing. For these reasons, the value of the special design stress of the material for the bearing is always very much below the design-stress values used in calculations of safe bearing loads on other machine parts, as, for example, in keys or flanged couplings. Values of the allowable bearing pressure range from a low of 35 psi to about 250 psi.

Illustrative Example. A 2-in.-diam shaft, 4 ft 0 in. long, supports a spur gear at a point 1 ft 6 in. from the right end. The combined weight and tooth thrust of the gear is 800 lb. Calculate the necessary length of bearing to support the shaft at the ends. The special design stress is 100 psi. Neglect the weight of the shaft.

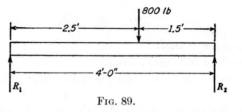

Fig. 89.

Refer to Fig. 89.

$$800 \times 2.5 - 4R_2 = 0$$

$$R_2 = \frac{800 \times 2.5}{4} = \frac{2000}{4} = 500 \text{ lb}$$

$$4R_1 - 800 \times 1.5 = 0$$

$$R_1 = \frac{800 \times 1.5}{4} = \frac{1200}{4} = 300 \text{ lb}$$

$$300 + 500 = 800 \text{ lb}$$

$$l = \frac{R}{sd} = \frac{500}{100 \times 2} = 2.5 \text{ in.} \tag{50}$$

80. Bushings. The oil film normally separating the moving shaft from its bearing, as mentioned in Sec. 79, is not present at starting and is destroyed at stopping. A certain amount of rubbing of metal on metal with attendant increased wear takes place at such times. Also in some cases the film at best is so thin that the tiny projections on the metal surfaces will pierce the film, rub against the companion piece, and again increase the wear.

As the wearing process continues and the shaft becomes loose in the bearing, replacements must be made. To keep the cost of this maintenance as low as possible, the hole of the bearing is originally made large enough to insert a collar called a bushing. The bushing, acting as a lining of the bearing, accommodates the shaft. Replacement of the bushing when worn is relatively easy and inexpensive. In order that the bushing will take the wear rather than the shaft (the replacement of which involves considerable expense), the bushing is made of a softer material such as brass, bronze, or babbitt metal (Chap. 3).

81. Friction in Bearings and Generated Heat. The problem of lessening frictional resistance is another factor in the choice of the most suitable metal for a bushing. Frictional resistance to sliding between two surfaces (as explained in Sec. 66) is equal to the force necessary to cause or maintain sliding at constant speed between the two contact surfaces.[1] Also we developed the equation

$$f = \frac{F}{N} \qquad (38)$$

An idea of the relative frictional resistances offered by a few of the more common materials when sliding on mild steel (both dry and clean) is obtained from the following tabulation.[2]

Material	Coefficient of Friction (f)
Cast iron	0.23
Babbitt metal	0.33
Brass	0.44
Nickel	0.64

Frictional resistance in sliding bearings is very much decreased when the revolving shaft is floating on a film of oil. In these cases frictional resistance depends on the sliding of oil molecules on oil molecules, that is, the internal friction of the oil itself.

Certain alloys such as bronze and the babbitt metals are chosen as bearing material for still another and very important reason. These alloys consist of a very fine mixture of one soft metal and a second that is considerably harder. When in use, the tiny particles of the softer metal wear down much more rapidly than the harder metal, leaving very small depressions on the surface. Oil collects in these pockets which then act as innumerable little reservoirs. In this way lubrication is aided at times of contact between bearing and shaft.

The heat generated at the bearing can now be calculated easily. Here the force pushing the stationary and sliding bodies together is the reaction of the shaft beam R. Hence Eq. (38) can be written

$$f = \frac{F}{R}$$

and

$$F = fR$$

For a rotating shaft, as discussed in Sec. 40,

$$\text{Work in 1 min} = P_1 \times \pi d \times n$$

[1] Frictional resistance of bodies at rest is somewhat greater than that of bodies in motion. For simplicity, this consideration is neglected in this text.

[2] From L. S. Marks, "Mechanical Engineers' Handbook," 5th ed., McGraw-Hill Book Company, Inc., New York, 1951.

To find the work lost in friction, P_1 is replaced by F or its equal fR. Then

$$\text{Work in 1 min} = fR \times \pi d \times n$$

Since d is in inches, we divide by 12 to get the answer in foot-pounds.

$$\text{Work in 1 min} = \frac{fR \times \pi d \times n}{12} \quad \text{ft-lb}$$

From physics we recall that

$$1 \text{ Btu} = 778 \text{ ft-lb of work}$$

Therefore,

$$\text{Btu per min} = \frac{fR \times \pi d \times n}{12 \times 778}$$

By substitution of a numerical value for π and cancellation, this expression reduces to

$$\text{Btu per min} = \frac{fRdn}{2970} \qquad (51)$$

Illustrative Example. Calculate the heat generated at the bearing for a 2½-in.-diam shaft revolving at a speed of 500 rpm. The weight of the shaft plus the weight and thrust of a pulley supported at the mid-section is 4500 lb. Assume the coefficient of friction of oil on oil as 0.03.

$$R = {}^{4500}\!/_2 = 2250 \text{ lb}$$

$$\text{Btu per min} = \frac{fRdn}{2970} = \frac{0.03 \times 2250 \times 2.5 \times 500}{2970} = 28.4 \quad (51)$$

82. Bearings with Rolling Contact. Consider the effort forces E and E' needed to push the equal weights W up the incline of Fig. 90a

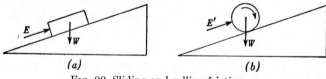

(a) (b)

Fig. 90. Sliding and rolling friction.

and b. Since in case a the weight must be slid up, whereas in case b it can be rolled up, there is no doubt that the necessary effort E for sliding is much more than E' for rolling. To put it another way, the coefficient of sliding friction is much greater than that of rolling friction. This statement is true, provided that the cross section of the roller is very nearly circular. If the surface is bumpy, as shown in Fig. 91, the effort E'' varies and at times may be more than E. This is an important fact in connection with ball and roller bearings. In contrast to the force of sliding friction, which depends on the load as well as the smoothness of the contact surfaces, the force of rolling friction shows very little increase with increased load. This means that when the shaft loads are large, ball or roller bearings are much more efficient than sliding-contact bearings. In ball bearings, values of f range between 0.0005 and 0.0030. For roller bearings, f is somewhat higher than 0.0030.

Fig. 91. Increased rolling friction.

Radial ball bearings are ball bearings in which the balls are placed in a ring around the shaft. Between balls and shaft there is a grooved ring or collar on which the balls roll (Fig. 92). Likewise between balls and support there is a similar grooved ring. These are called the inner race and outer race, respectively. The inner race is held to the shaft either by a very close fit or, when necessary, by a clamping device (not shown in the figure), to prevent sliding between these two parts. Also the outer race should not slide on the inside of the housing.

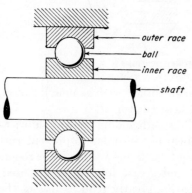

Fig. 92. Radial ball bearing.

Figure 93a and b are pictorial representations of single-ring ball- and roller-bearing assemblies. Here in each case is shown the bearing *retainer* also called the *cage* (omitted in Fig. 92), the purpose of which is to keep the balls or rollers from contact with one another and thus prevent wear. For roller bearings, holes in the retainers serve as bearings for the axles of the roller wheels. In double-ring bearings, there are two sets of balls or rollers between inner and outer race, each set held by its own cage.

Several factors should be considered in the selection of the material for the rollers. The first consideration is the matter of compressive stress. The radii of inner and outer race for ball bearings are somewhat larger than the radius of the ball. Theoretically, this would give a point contact between ball and race, but the elastic deformation of the ball under load results in a certain amount of flattening. Nevertheless, the contact area is very small, and this means a high compressive stress (pounds per square inch) over this area even for moderate shaft loads. This factor is less important in roller bearings. Why? Second, because of the repetition of stress in the ball as the shaft rotates and the ball rolls, fatigue failure is a distinct possibility. Such failure results in a flaking of the ball's surface. A third consideration is the necessity for a high degree of hardness to minimize wear and save replacement of the balls. It has been found that a special alloy steel

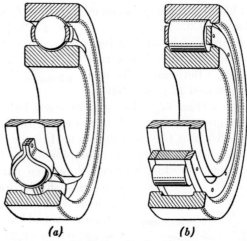

(a)   *(b)*

Fig. 93. Ball and roller bearings.

with alloying elements of nickel and some other, such as molybdenum, is best. This steel is given a special heat-treatment.

While not as important as in bearings with sliding contact, lubrication is necessary for ball and roller bearings in order to prevent rust and for the small amount of sliding that exists between balls or rollers and race as well as cage. Both oil and grease lubrication are used. A housing surrounding the entire bearing greatly reduces loss of the lubricant.

83. Thrust Bearings. Our discussion of bearings thus far has been limited to cases where the force of the shaft is perpendicular to the shaft itself. In many cases, however, the major force runs parallel to the shaft, as we shall see. The rotation of the shaft and worm of Fig. 94 drives the gear. But the gear offers resistance to being driven and in turn tends to push the worm and shaft to the left, as indicated by the arrows. The bearing must withstand this thrust and keep the shaft in place. This is called a *thrust bearing*.

A *collar thrust bearing* is one in which the axial force is transmitted through collars fitted to the shaft, as illustrated in Fig. 95. Each collar transmits an equal share of the axial push and the sum of all the collar areas multiplied by the bearing stress equals the thrust.

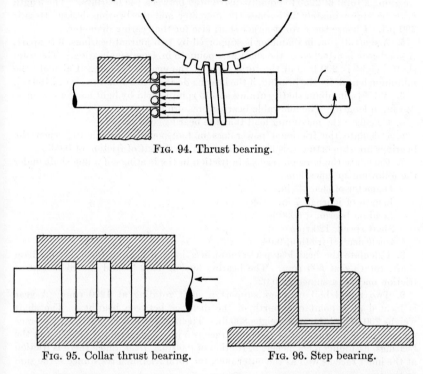

FIG. 94. Thrust bearing.

FIG. 95. Collar thrust bearing.   FIG. 96. Step bearing.

Figure 96 illustrates the cross section of a type of thrust bearing called a *step bearing* where the thrust is caused by the weight of the shaft and its gears or other attachments. When balls are not used, one or more thin disks may be placed between bearing and end of shaft. The disks will revolve at speeds lower than the shaft and take up some but not all of the wear. They are easily replaceable.

PROBLEMS

1. The combined weight of a 2-in.-diam shaft and pulley and the belt tensions create a maximum bearing reaction of 2350 lb. Calculate the necessary length of the bearing, if the special design stress for the bearing is 220 psi.

2. A journal bearing for a $2\frac{1}{4}$-in.-diam shaft rotating at 200 rpm is subjected to a load of 1800 lb. The ratio of the length to diameter is 3. Calculate:

a. The projected area

b. The pressure at the bearing in pounds per square inch of projected area

3. The maximum allowable bearing pressure on a 2-in.-diam bearing is 80 psi. The reaction on the bearing from the shaft is 500 lb. Make calculations to determine whether a bearing length of $3\frac{1}{2}$ in. is sufficient.

4. Calculate the dimensions for a journal bearing to support a shaft and flywheel weighing a total of 2250 lb equally distributed between the bearings. The length is to be approximately $2\frac{1}{2}$ times the diameter and the bearing design stress is 200 psi. Choose the nearest larger shaft size for the bearing diameter.

5. A shaft $3\frac{1}{2}$ in. in diameter is supported by two journal bearings, 6 ft apart. A spur gear is attached to the shaft at a point 15 in. from one bearing. The total shaft weight is 200 lb and the total force of the gear is 3500 lb. Calculate the minimum length for the bearing, if the bearing design stress is not to exceed 160 psi.

6. A $1\frac{15}{16}$-in.-diam shaft, turning at 150 rpm, is held by light hanger bearings, $2\frac{1}{2}$ in. in length. The allowable bearing pressure is 40 psi.

  a. Calculate the maximum load that the bearings can carry.

  b. Calculate the frictional power loss in foot-pounds per minute, when the bearings are supporting this load. Assume a coefficient of friction of 0.035.

7. Calculate the horsepower lost in friction in the bearings of a line shaft under the following specifications:

  Diameter of shaft, 3 in.
  Length of journal, 7 in.
  Load on bearings, 6000 lb
  Shaft speed, 120 rpm
  Coefficient of friction, 0.04

8. Calculate the heat lost per minute in a journal bearing for a $1\frac{1}{2}$-in.-diam shaft, rotating at 500 rpm. The bearing load is 1100 lb and the coefficient of friction may be assumed as 0.025.

9. Two radial ball bearings support a shaft rotating at 1200 rpm. A gear is located at a point one-fourth of the distance between bearings. The shaft weighs 200 lb and the gear weighs 800 lb. The tooth thrust is 950 lb. Inner-race diameter is 3 in. and that of the outer race is 5 in. Calculate the heat lost in the bearings, assuming a coefficient of friction of 0.0015. *Hint:* Since there is friction at the inner race as well as the outer race, the total frictional resistance is the sum of the two.

10. Calculate the proper length of bearing for a 5-in.-diam marine propeller shaft designed to transmit 150 hp at 90 rpm. The maximum bearing load is 15,000 lb. The allowable bearing pressure is 200 psi and the allowable amount of heat which may be generated is 0.85 Btu per min per sq in. of projected area. Assume a coefficient of friction of 0.02.

11. Four collars are to be fitted to the end of a 3-in.-diam shaft to transmit a thrust of 4000 lb. The allowable bearing pressure on the collars is 70 psi. Calculate the necessary outside diameter of the collars.

12. A step bearing supports a $2\frac{1}{2}$-in.-diam shaft. A disk with a $\frac{3}{4}$-in. hole at its center is placed between shaft and bearing. The pressure on the bearing is 55 psi. Calculate the heat generated in Btu per minute when the shaft revolves at a speed of 1200 rpm. Assume a coefficient of friction of 0.03. *Hint:* For simplicity, d in formula (51) may be assumed as the shaft diameter, that is, $2\frac{1}{2}$ in.

CHAPTER 16

CLUTCHES

84. The Clutch Function. If an operator of an automobile pushes the clutch pedal downward to disengage the clutch, the motor can no longer propel the car and its velocity rapidly decreases. By his action, the operator has freed the motor from the driving mechanism. Clutches are used in various types of machines when it is necessary to change the direction of motion or speed of the machine by shifting the gears, which operation cannot be performed while the motor is engaged to the drive shaft. To eliminate the bother of gear shifting, the fluid-drive mechanism has within recent years found favor among manufacturers of automobiles. Actually, in some types the fluid drive does not eliminate the shifting of gears but makes this action automatic. In a recent development, the fluid clutch is given magnetic properties, thereby lessening greatly the power losses in starting.

The most important consideration in clutch design is to transmit the power through the clutch engagement without shock. Therefore, all clutches (excepting those attached to shafts revolving very slowly) must gradually engage the motor to the driving mechanism.

Because of the relative massiveness of the clutch and the fact that it protrudes a considerable distance from the axis of rotation, as well as its symmetry about that axis, the clutch has the qualifications of a flywheel and in fact is frequently used to fulfill the flywheel function also.

85. Friction Clutches. For some years the automobile industry has favored the use of the disk friction clutch. The operational principle of this device is illustrated in Fig. 97. The thin driven plate slides on a splined portion of the main shaft; the flywheel is keyed fast to the crankshaft. Power is transmitted by means of the high frictional forces developed when the compression springs push the pressure ring against the driven plate, which in turn pushes against the flywheel. Thus the friction on both sides of the driven plate is utilized. A special "throwout" mechanism (not shown here) removes the forces of the compression springs when the motor is to be disengaged. In the engaging operation, the spring force should be applied gradually to prevent shock.

Another widely used type of friction clutch is the cone clutch (Fig. 98). In this case the compression spring exerts the force P_a axially

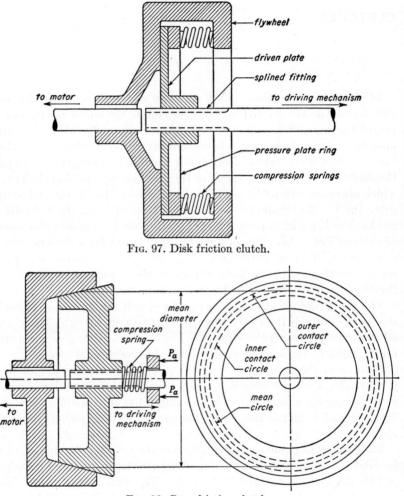

FIG. 97. Disk friction clutch.

FIG. 98. Cone friction clutch.

to press the inner cone firmly into the outer one. Again, power transmission depends on the friction developed between the two contact surfaces. Pulling the inner cone to the right breaks the contact and, hence, disengages motor shaft and drive shaft. The outer cone (usually the driver) is keyed fast to the shaft and the inner cone slides on a spline.   The throwout mechanism is not shown in the figure.

86. Design Theory of the Cone Friction Clutch.   In order to arrive at a clearer understanding of the action of the forces involved in the

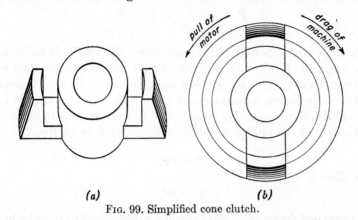

(a)   (b)

FIG. 99. Simplified cone clutch.

cone clutch, let us imagine a clutch with most of the surface of the inner cone removed.   All that remains of the original lateral surface of the cone are the two small parts opposite to each other, as represented in Fig. 99a and b by the shaded areas.   This reduced inner cone, when functioning in a clutch is similar to a steel wedge driven into, let us say, a piece of wood.   The forces acting, as shown in Fig. 100, are the axial thrust P_a of the spring and the push $P_n/2$ of the outer cone, perpendicular (normal) to the slanting surface at the top, and its counterpart at the bottom.   Although for simplicity the forces $P_n/2$ are shown as acting at a point on the mean circle, it is to be remembered that they act over the entire shaded areas of Fig. 99a and b which areas must exert sufficient frictional resistance to

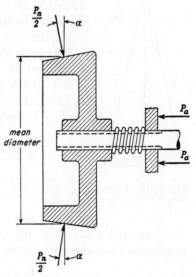

FIG. 100. Forces on inner cone of cone clutch.

prevent slippage with the surface of the outer cone (see Fig. 99b). The frictional resistance F, as we discussed in Sec. 66, is the product of the coefficient of friction f and the force normal to the surfaces which presses these surfaces together, $P_n/2$ in this case. Then the entire frictional resistance for both surfaces would be

$$F = f \times \frac{P_n}{2} \times 2 = fP_n$$

After an initial period of slippage when the clutch is being gradually engaged, no further slippage should take place between inner and outer cones. This means that the clutch must be able to develop a frictional resisting moment equal to the driving torque. We know that the frictional resisting moment is the product of the frictional resistance and the radius of the mean circle, and, in symbols,

$$T = Fr$$

Also, since $F = fP_n$,

$$T = fP_n r$$

The value of f depends on the kind of materials involved and r in any particular clutch is known or determined from measurement. Therefore, P_n remains as the only unknown in the determination of T. We shall now see how P_n can be expressed in terms of P_a, again a known value in any particular clutch.

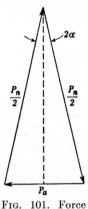

Referring again to Fig. 100, we note that $P_n/2$, $P_n/2$, and P_a hold the clutch in equilibrium, and hence, as the directions of these forces are known, $P_n/2$ can be found in terms of P_a. From the fact that for equilibrium $\Sigma H = 0$,

$$\frac{P_n}{2} \times \sin \alpha + \frac{P_n}{2} \times \sin \alpha - P_a = 0$$

$$P_n \sin \alpha - P_a = 0$$

$$P_n = \frac{P_a}{\sin \alpha}$$

Fig. 101. Force polygon for cone clutch.

The force polygon of Fig. 101 verifies this relationship. Drop a perpendicular from the apex to the base of the triangle. Then

$$\frac{P_a}{2} = \frac{P_n}{2} \times \sin \alpha$$

and

$$P_n = \frac{P_a}{\sin \alpha}$$

We are now able to express the torque T in terms of all known values. As previously developed, $T = fP_n r$, which by substitution of $P_a/\sin \alpha$ for P_n becomes

$$T = \frac{fP_a r}{\sin \alpha}$$

In order to calculate the horsepower that a clutch can deliver, let us refer back to our fundamental horsepower expression as developed in Sec. 40, namely,

$$H = \frac{Tn}{63,000} \tag{11}$$

where H is the power in horsepower.

By substituting the value of T just found, this expression becomes

$$H = \frac{fP_a r}{\sin \alpha} \times \frac{n}{63,000}$$

and

$$H = \frac{fP_a rn}{63,000 \sin \alpha} \tag{52}$$

This equation can also be used to calculate the required spring force, when the horsepower and speed are known, for

$$P_a = \frac{63,000H \sin \alpha}{frn}$$

Furthermore, since from Eq. (11) $T = 63,000H/n$, T can be substituted for its equal in Eq. (52) to obtain

$$P_a = \frac{T \sin \alpha}{fr} \tag{53}$$

Equation (53) is used when the torque to be delivered is known.

It should be recalled at this time that, in order to understand the action of the forces more clearly, we assumed that the lateral surface of the inner cone was cut away to the extent shown in Fig. 99a and b, thus leaving intact the two small opposite parts only of this lateral surface. When the axial force is applied, nothing more than these parts would be in contact with the outer cone. Although the inner cones of friction clutches are not cut away in this fashion and do have a continuous lateral surface as indicated in Fig. 98, the axial force P_a does cause the same forces $P_n/2$ and $P_n/2$ for equilibrium, with the exception that, instead of two forces, there are an infinite number of forces all normal to the lateral surface with the sum of all of them equal to P_n. Our assumption was, therefore, valid and Eqs. (52) and (53) hold true.

Illustrative Example. A flywheel cone clutch transmits a twisting moment of 1800 lb-in. The coefficient of friction between inner and outer cone is 0.35, the mean diameter is 20 in., and the clutch angle is 9°30′. Calculate the necessary force of the spring.

$$P_a = \frac{T \sin \alpha}{fr} = \frac{1800 \times \sin 9°30′}{0.35 \times 10} = \frac{1800 \times 0.1650}{0.35 \times 10} \quad (53)$$
$$P_a = 84.9 \text{ lb}$$

The material most suitable for inner and outer cones is cast iron. For the contact surfaces, a number of factors enter into the selection of the material. As previously stated, the foremost consideration in clutch design is transmission of power without shock. In accomplishing this, a large amount of slippage between contact surfaces is present during the engaging operation, resulting in considerable wear and heat generation. The material for the surfaces, therefore, should possess the ability to resist wear and not disintegrate at elevated temperatures. The latter aspect is especially important for clutch mechanisms in more or less continual use, as in buses driven in city traffic. Furthermore, to reduce the necessary spring force as well as to keep down wear and temperature, the material should have a high coefficient of friction. Asbestos fabric meets these requirements and is widely used. Sometimes wood or leather or both are used. When worn, cone coverings can be fairly inexpensively replaced. In cases where no coverings are used, that is, where there is metal-to-metal contact, excessive wear means costly replacements. Such surfaces should be kept oiled to reduce wear.

To reduce the amount of heat generated and to minimize wear per square inch of contact surfaces, it is important that the normal pressure (in pounds per square inch) be kept low. This means that the contact area should be of sufficient size to distribute the normal load P_n for

$$p_n = \frac{P_n}{A}$$

where p_n is the normal pressure in pounds per square inch. Values of p_n may range from 15 to 30 psi.

Illustrative Example. Calculate the width of the cone contact surfaces necessary in a cone clutch to transmit 20 hp at 300 rpm. The mean radius of the cone is 5 in., the clutch angle is 10°, the normal pressure on the cone surfaces should not exceed 25 psi, and the coefficient of friction of the contact materials is 0.5.

$$P_a = \frac{63,000H \sin \alpha}{frn} = \frac{63,000 \times 20 \times 0.1736}{0.5 \times 5 \times 300} = 292 \text{ lb} \quad (52)$$

$$P_n = \frac{P_a}{\sin \alpha} = \frac{292}{0.1736} = 1680 \text{ lb}$$

$$A = \frac{P_n}{p_n} = \frac{1680}{25} = 67.2 \text{ sq in.}$$

Also

$$A = w \times \pi d$$

$$w = \frac{A}{\pi d} = \frac{67.2}{3.14 \times 10} = 2.14 \text{ in.}$$

The selection of the value of the clutch angle α is another important consideration in design. If too small, the wedge action of the inner cone is increased and great force is needed to pull the cones apart. On the other hand, too large an angle means that the forces making up P_n become small, and, since this means the friction is reduced, there may not be sufficient frictional resistance developed to transmit the required

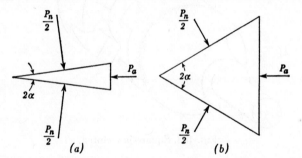

FIG. 102. Effect of increasing the clutch angle in the cone clutch.

power. These two conditions are illustrated diagrammatically in Fig. 102a and b. For equal values of P_a, the forces making up $P_n(P_n/2$ and $P_n/2$ in this case) necessary to maintain equilibrium become larger as the angle 2α becomes smaller. This fact can be verified by drawing the force polygons for conditions a and b of Fig. 102. A value of $12\frac{1}{2}°$ for α is recommended by the Society of Automotive Engineers (SAE).

Owing to the fact that α is always a small angle, a small amount of wear on the cone surfaces results in a comparatively large increase in the distance that the inner cone must move to engage the clutch. Hence, the spring force P_a and the normal force P_n are reduced. Provision is made in some clutches to keep the spring force constant, thus eliminating this effect of wear.

As previously mentioned, the clutch frequently serves also as a flywheel. However, machines equipped with large and heavy flywheels (according to the kinetic energy equation developed in Sec. 30) are difficult to stop, and therefore massive clutches should be avoided in high-speed machinery where frequent braking action is necessary.

Much useful information can be gathered from clutch manufacturers' catalogues. Many of them include tables or graphs to show the horsepower ratings for various clutch sizes, to aid the prospective buyer in making proper selections.

87. Positive Clutches. In a low-speed mechanism the effect of shock is minimized to permit the use of clutches which do not allow

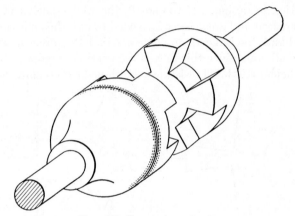

Fig. 103. Square-jaw clutch.

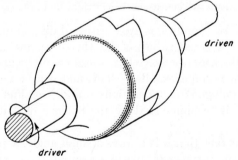

driven

driver

Fig. 104. Spiral-jaw clutch.

any slippage during engagement. These are called *positive clutches*, two types of which are shown in Figs. 103 and 104. As in other types of clutches described above, one half is keyed fast to one shaft, while the other half is free to slide along the other colinear shaft on a spline.

In the *square-jaw clutch* of Fig. 103, the length of the lugs should be about 20 per cent less than the length of the sockets in order that the lugs will slip easily into position when the clutch is being engaged and one half is in rotation. The *spiral-jaw clutch* of Fig. 104 engages more easily than does the square-jaw clutch. For opposite rotation, the spirals must run in the opposite direction.

PROBLEMS

1. The force of each of four compression springs of a disk clutch is 150 lb. What is the minimum distance that these springs can be placed from the shaft center in order to transmit 90 hp at 1400 rpm. Use a coefficient of friction for metal on metal of 0.25.

2. Calculate the axial spring force for a cone friction clutch necessary to transmit 50 hp at a speed of 1250 rpm. The clutch angle is 12°30′, the mean cone diameter is 12 in., and the coefficient of friction of the rubbing surface is 0.45.

3. A cone clutch must transmit the torque from a 26-in.-diam gear with a tooth thrust of 300 lb. The mean diameter of the clutch is 14 in. and the clutch angle is 10°. Calculate the axial force, assuming a coefficient of friction of 0.60.

4. What should be the minimum mean diameter of a cone clutch to transmit 30 hp at 800 rpm with a spring force of 175 lb? Assume a clutch angle of 12°30′ and a coefficient of friction of 0.55.

5. The normal force on the cones of a cone friction clutch is 4000 lb. The width of contact faces is 3 in. and the mean diameter is 20 in. Calculate the normal pressure.

6. To prevent excessive heat, the normal pressure on a cone clutch is limited to 25 psi. The mean cone diameter is 18 in.

a. Calculate the necessary width of the contact faces for a normal force of 2500 lb.

b. What is the clutch angle, if the axial force on the cones is 300 lb?

7. A cone clutch is to be used to transmit a torque of 2600 lb-in. The mean diameter is 10 in., the clutch angle is 12°30′, the coefficient of friction is 0.25, and the normal pressure must not exceed 20 psi. Calculate the necessary width of contact faces.

CHAPTER 17

THIN-WALLED CYLINDERS;
WELDED AND RIVETED JOINTS

88. Boilers and Pipes. The design of boilers is a very important branch of machine design. As advances in industry have required the use of ever-increasing steam pressure, the designer has been called upon to increase boiler strength and at the same time keep down the cost of construction as much as possible. Today boilers are built to withstand a steam pressure of as much as 2000 psi.

Boilers and most pipes have thin walls in comparison to their diameters. The stresses in such vessels lend themselves to a simple analysis, and, hence, the vessels are easy to design. In this chapter we are concerned with the design of such vessels—their walls and joints.

89. Design of Thin-walled Cylinders. Let us imagine that the lower half of the cross section of the boiler of Fig. 105 is filled with a solid

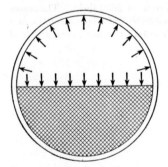

FIG. 105. Effective boiler pressure on sides.

substance such as concrete. Then steam under pressure is allowed to enter. The arrows indicate the pressure on the inside wall and on the concrete in each case normal to the surface. With the exception of the arrow at the extreme top, all arrows on the wall have both sidewise and upward components. It is a fact that the total upward thrust for any length of boiler must be equal to the downward thrust on the concrete for the same length. If it were not so, the steam pressure would cause motion of the boiler up or down. To determine the total upward or downward force on any length, therefore, we need merely calculate the force on the flat surface of the concrete. In Fig. 106 the force on the flat surface is the pressure p in pounds per square inch multiplied by the number of square inches in that surface, namely,

$$P = p \times l \times D$$

This force is resisted by a stress of tension at this longitudinal section with an area of $2 \times t \times l$. The total force that the area can with-

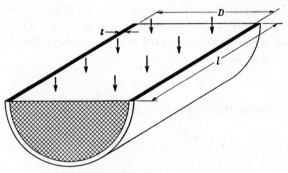

FIG. 106. Resisting area to side pressure.

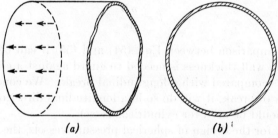

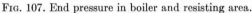

(a)   *(b)*

FIG. 107. End pressure in boiler and resisting area.

stand safely is that area multiplied by the design stress of the material, or

$$P = 2 \times t \times l \times s$$

Then by equating the two values of P we have

$$p \times l \times D = 2 \times t \times l \times s$$

which by cancellation of l and transformation becomes

$$t = \frac{pD}{2s} \tag{54}$$

From this equation the required thickness of the boiler can be determined.

The inside pressure also causes a force on the ends of the boiler. This force (Fig. 107a) is the pressure multiplied by the area of the end, that is,

$$P = p \times \frac{\pi D^2}{4}$$

The resisting area in this case is the area of the circle of Fig. 107b. Strictly speaking, this area is the difference between the areas of the outside and inside circles, but for simplicity, we take the inside circumference and multiply it by the thickness: $t \times \pi \times D$. This area multiplied by the stress equals the load and, for safety, the design stress must not be exceeded. Therefore,

$$P = t \times \pi \times D \times s$$

Again, by equating the two values of P, we have

$$p \times \frac{\pi D^2}{4} = t \times \pi \times D \times s$$

which by cancellation and transformation becomes

$$t = \frac{pD}{4s} \tag{55}$$

From a comparison between Eqs. (54) and (55) it can be seen that only half the wall thickness is needed to guard against a circumferential break as compared with a longitudinal break. We conclude that, if a boiler is to break, it will do so in a longitudinal direction. Equation (54) should be used for cylindrical vessels.

However, for the design of spherical pressure vessels, the total force that must be resisted by the material is the same as the force on the *end* of a cylindrical vessel of equal diameter. Also, as in the case of a cylinder, the material at the cross section is resisting this force. For spheres, Eq. (55) applies.

90. Joints of Pressure Vessels—Efficiency. If it were not for the joints in pressure vessels, no further design would be required, but unfortunately joints are necessary. These are either welded or riveted, with the welded joints assuming the greater importance in recent years.

No joint either welded or riveted is considered to be as strong as the solid plate. In welds, among other things, there may be an annealing effect in the higher carbon steels which softens the material around the joint and makes it less strong. In riveted joints it is obvious that the rivet holes weaken the plate. As developed in Sec. 66, the ratio, expressed as a percentage, of the strength of the joint to the strength of the solid plate is called the efficiency of the joint.

Mathematically,

$$\eta = \frac{\text{strength of joint}}{\text{strength of solid plate}} \tag{37}$$

where η stands for efficiency. Note that the efficiency for a boiler or pipe joint is never as much as 100 per cent.

A pressure vessel is no better than its joint, and therefore the joint must be made sufficiently strong to withstand the pressure. This means that the solid plate will be made somewhat stronger than necessary by increasing its thickness. The lower the efficiency, the thicker the plate must become. The following proportion holds true.

$$\frac{t}{t'} = \frac{\text{strength of joint}}{\text{strength of solid plate}} = \eta$$

where t and t' are the original and increased wall thicknesses, respectively. Also

$$t' = \frac{t}{\eta}$$

but

$$t = \frac{pD}{2s} \tag{54}$$

Therefore,

$$t' = \frac{pD}{2s\eta} \tag{56}$$

For welded joints, the efficiency may be as high as 90 or even 95 per cent. For riveted joints, the average value is much lower.

Illustrative Example. Calculate the required wall thickness for a boiler, 6 ft 0 in. ID and 10 ft 6 in. long, to withstand a steam pressure of 300 psi. Assume an efficiency of 90 per cent for the joints and a design stress of 10,000 psi for the steel in tension.

$$t' = \frac{pD}{2s\eta} = \frac{300 \times 72}{2 \times 10,000 \times 0.90} = 1.20 \text{ in.} \tag{56}$$

Boiler plate comes in thicknesses of $\frac{1}{16}$-in. increments. Use the next regular thickness of $1\frac{1}{4}$ in., or 1.25 in. In problems of this type be sure to express the diameter in inches before substituting in the formula.

91. Welded Joints. The most common type of welding is fusion welding. This process consists of heating the contact surfaces of the joint to the temperature at which the metal becomes very soft. In most cases new metal (as from a rod) of the same kind of metal is heated and added to the joint. Then upon solidification the joint becomes a fairly homogeneous mass. Heat is often applied by the burning of two gases, such as oxygen and acetylene.

Figure 108 shows a few of the many types of welds. Fillet welds are made between faces or the edge of one plate and the face of another (Fig. 108a). Butt welds join the ends or edges of plates (Fig. 108b).

The strength of a butt weld may be calculated by assigning a design stress to the weld and multiplying by the cross-sectional area. Although the cross section at the joint may be somewhat larger than the cross section of the plate (see Fig. 108b), the plate cross section is

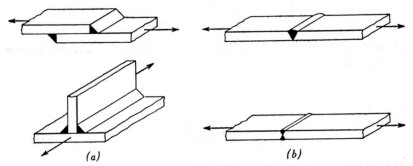

(a)   (b)

FIG. 108. Types of welded joints. (a) Fillet welds; (b) butt welds, single-V and double-V.

used in this calculation. More frequently an efficiency is assigned to the weld and this value is multiplied by the strength of the solid plate to obtain the strength of the joint.

The design-stress values recommended for welded joints vary with the type of loading. As in other machine-design applications, repeated loads, reversal of loads, and suddenly applied loads, all mean a lowered design-stress value. For tension (usually butt welds) values vary from 6000 to 12,000 psi, depending on the type of loading. Compression values are somewhat higher. For shear (fillet welds) design-stress values from 4000 to 8000 psi are common.

Illustrative Example. Calculate the strength of the lap fillet-welded joint shown in Fig. 108a (top). The base of weld is $\frac{1}{2}$ in. and the width of plate is 14 in. Use a design-stress value of 6000 psi.

$$P = As = 0.5 \times 14 \times 2 \times 6000 = 84,000 \text{ lb} \qquad (1)$$

92. Riveted Joints. Riveted joints are of two general types: lap joints and butt joints. Examine Fig. 109.

The ways in which riveted joints may fail are as follows:

a. Shear of the rivets

b. Crushing of the plate metal by the rivets, known also as bearing failure

c. Tension in the plate

d. Shear in the plate

e. Tearing of the plate

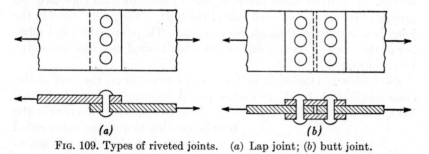

FIG. 109. Types of riveted joints.   (*a*) Lap joint; (*b*) butt joint.

These causes of failure are illustrated in Fig. 110 for a single-rivet lap joint. They apply to butt joints as well. The side view is shown in *a* to illustrate the severing of the rivet by shear. In *d* and *e* the plate can be strengthened against these failure possibilities by increasing the margin or edge distance. If this distance is made no less than

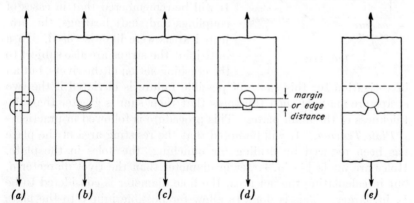

FIG. 110. Failure possibilities in riveted joints.   (*a*) Rivet shear; (*b*) plate bearing; (*c*) plate tension; (*d*) plate shear; (*e*) plate tearing.

$1\frac{1}{2}$ times the rivet diameter, there will be no danger of failure from sources *d* and *e*. Therefore, in our calculations of riveted joints, we shall consider the first three types of failure only.

Since the induced stress involved in each type of riveted joint failure is a direct stress (shear, bearing, or tension), the load causing failure in each case is found by multiplying the area by the stress, namely,

$$P = As \qquad (1)$$

When the design stress is used, the safe load is obtained. It is unnecessary to learn specific formulas, and such formulas will not be presented. It is to be noted carefully that in the butt joint there are two groups of rivets, one on each side of the butt. The *entire* stress in the joint is transmitted through each group. Therefore, the total number of rivets *either* on one side or the other is used, not the sum of the two groups.

Rivet Shear. Obviously in the case of rivet shear, the sum of the cross sections of the rivets makes up the resisting area. For the lap joint, one area in each rivet resists the shearing force, whereas in the butt joint with two straps (also called cover plates) each rivet has two resisting areas (refer to Fig. 109).

Plate Bearing. The metal against which the rivet pushes or bears must resist the push. Instead of calculating the area of the semicircular surface, the projection of this area is used (Fig. 111). It will be remembered that in cases of couplings and shaft bearings, the projected areas were likewise used. In a butt joint the straps are also subject to the crushing action of the rivets, but an investigation for this cause of possible failure is obviated if the two straps are made of such thickness that their sum is greater than the thickness of the main plate. This procedure is followed in practice.

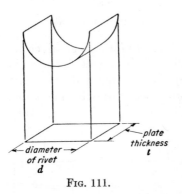

plate thickness *t*

diameter of rivet *d*

Fig. 111.

Plate Tension. It will be noted that the resisting area of the plate has been reduced by drilling (or punching) the holes in the plate. Holes are made $\frac{1}{16}$ in. larger in diameter than the rivet diameter *d*, but in calculating the net area, the hole diameter is considered to be $\frac{1}{8}$ in. larger. This is done to allow for possible injury to the plate where the hole is made. For each hole the area is reduced by the product of $d + \frac{1}{8}$ and the plate thickness *t* (or *t'*).

A joint is no stronger than its weakest part. Thus in calculating efficiency of a riveted joint, care must be taken to select the lowest value of the safe load calculated as the strength of the joint. This value may be that of rivet shear, plate bearing, or plate tension.   Table 8 gives average values of lap- and butt-joint efficiencies. Although more costly, butt joints have greater efficiencies.

For riveted joints, as in welded joints, lower design stress values are used for machine-design work than is common in structural work. In our problems we shall assume values of design stresses as follows: shear, 8500 psi; bearing, 20,000 psi; and tension, 10,000 psi.

TABLE 8. LAP- AND BUTT-JOINT EFFICIENCIES

Type of Joint	Average Efficiency, Per Cent
Lap:	
Single row	55
Double row	68
Triple row	75
Butt:	
Single row	67
Double row	80
Triple row	83
Quadruple row	91

Illustrative Example 1.   A riveted lap joint consists of three 1-in. rivets connecting two plates each ½ in. thick and 9 in. wide.   The rivets are in one row parallel to the plate edges.

a. Calculate the safe tensile load that can be applied to the joint.

b. Calculate the efficiency.

a. Strength of joint:
   Shear in rivets:

$$P = As = 3 \times 0.785 \times 1^2 \times 8500 = 20,000 \text{ lb} \quad (1)$$

Bearing in plate:

$$P = As = 3 \times 0.5 \times 1 \times 20,000 = 30,000 \text{ lb} \quad (1)$$

Tension in plate:

$$P = As = [9 - 3(1 + 0.125)]0.5 \times 10,000 = 28,100 \text{ lb} \quad (1)$$

The strength of the joint is 20,000 lb.

b. Efficiency of joint:

$$\eta = \frac{\text{strength of joint}}{\text{strength of solid plate}} \quad (37)$$

Strength of solid plate = $9 \times 0.5 \times 10,000 = 45,000$ lb

$$\eta = \frac{20,000}{45,000} = 44.4\%$$

This value is very low.

Illustrative Example 2.   Calculate the safe load and the efficiency for the riveted butt joint of Fig. 112.

Shear in rivets:

$$P = As = 7 \times 0.785 \times 0.875^2 \times 2 \times 8500 = 71,500 \text{ lb} \quad (1)$$

Bearing in plate:

$$P = As = 7 \times 0.625 \times 0.875 \times 20,000 = 76,600 \text{ lb} \quad (1)$$

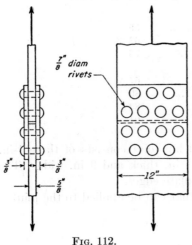

FIG. 112.

In the case of tension in the plate, the first thought is to investigate the section of smaller net area, which is the one through the centers of the four holes. However, if the main plate is to break at this section, the three-hole section adjacent must also fail before the joint will sever. This second failure might be in rivet shear or plate bearing. On the other hand, the plate might fail in tension at the three-hole section without affecting the four-hole section. Therefore, the three-hole section will be investigated.

Tension in plate:

$$P = As = [12 - 3(0.875 + 0.125)]0.625 \times 10,000 = 56,200 \text{ lb} \quad (1)$$

Strength of solid plate:

$$P = As = 12 \times 0.625 \times 10,000 = 75,000 \text{ lb} \quad (1)$$

$$\eta = \frac{\text{strength of joint}}{\text{strength of solid plate}} = \frac{56,200}{75,000} = 75\% \quad (37)$$

93. Riveted Joints in Boilers.   In boiler construction today, welded joints are used mostly, but riveted joints persist. The butt joint illustrated in Fig. 113 is typical for boilers. Note that the inside strap is used for three rows of rivets but the outside strap for only two. The rivets run along the joint in a regular repeating pattern. One portion with a complete pattern is known as a *repeating section*, as indicated in the figure. Since each of these repeating sections is alike, one only need be designed.

The distance between rivet centers in a single row is called the *pitch, p.* In the illustration we have two such distances, p_1 and p_2. A line passing through a row of rivets is a *gage line* and the distance between gage lines is the *transverse pitch, p_t.* When rivets are not directly behind each other in each gage line they are said to be *staggered.* For staggered rivets the distance from the center of one rivet to the center of the nearest one on the next gage line is called the *diagonal pitch, p_d.*

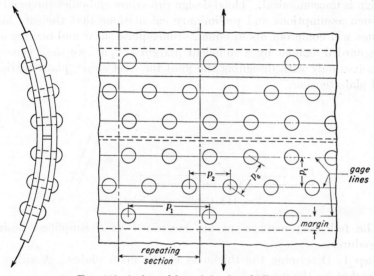

FIG. 113. A riveted butt joint in a boiler.

Rivet pitch is made a minimum of $3d$ and of such a maximum distance that the vessel can be made leakproof for its containing fluid.   To ensure sufficient room for the riveting tools, the transverse pitch must not be less than $2.5d$.   This minimum distance also means that the diagonal pitch will be long enough to prevent tension failure along this line.   The margin should be a minimum of $1.5d$.

A process known as *caulking* is applied to the edges of the plates in a lap joint and the strap in a butt joint.   These edges are first beveled to approximately 75° and the caulking tool hammered on the edge by a hand or power hammer.   The purpose of caulking is to make joints leakproof (refer to Fig. 114).

The material selected for steam boilers is almost exclusively plain carbon steel with a low carbon content (AISI 1010 or 1020).   The great toughness and good strength of this material are important factors.   In certain industries where the substance to be contained in tanks and pressure vessels may react with the steel, other materials are chosen, as, for example, the use of copper in breweries.

94. Design of a Riveted-joint Boiler. In any riveted joint it is desirable to have the safe loads equal as calculated from each possible way of failure. If, for example, one investigation results in a considerably greater safe load than the others, the joint is needlessly strong in this respect. On the other hand, if any one safe load value is much below the others, the efficiency will be reduced greatly and again the design is uneconomical. Good design procedure embodies the making of such assumptions and preliminary calculations that the safe load values will come out about equal. Since plate shear and bearing are safeguarded against by a sufficient margin ($1\frac{1}{2}\,d$), we shall concern ourselves only with designing the joint for rivet shear, plate bearing, and plate tension.

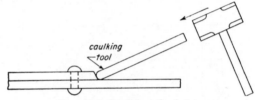

Fig. 114. Caulking the joint.

The following main steps are suggested as a simplified design procedure.

Step 1. Determine the thickness of the main plates. A value is arrived at by the use of

$$t' = \frac{pD}{2s\eta} \tag{56}$$

An efficiency is assumed.

Step. 2. Select the rivet diameter d, usually by a rule-of-thumb method, and calculate miscellaneous dimensions.

Step 3. Determine the pitch. This is done by equating the safe loads for shear and tension and solving for p.

Step 4. Calculate the actual safe loads for rivet shear, plate bearing, and plate tension.

Step 5. Calculate the efficiency and check with the assumed value. A redesign may then be necessary.

The simple example given below will help to make the procedure more clear.[1]

---

[1] In order to increase joint efficiency, the rivet pattern is often made quite complicated. The design of such layouts becomes quite involved. For our purposes the simple illustration given will suffice.

Illustrative Example. A boiler, 4 ft 0 in. in diameter, must withstand a steam pressure of 175 psi. Design the wall thickness and a longitudinal joint according to the rivet layout of Fig. 115.

1. From Table 8 we assume an efficiency of 80 per cent.

$$t' = \frac{pD}{2s\eta} = \frac{175 \times 48}{2 \times 10,000 \times 0.80} = 0.525 \text{ in.} \qquad (56)$$

Use $\frac{9}{16}$-in. plate = 0.563 in.

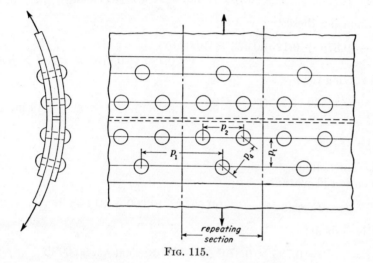

Fig. 115.

2. In a butt joint with moderate plate thickness, the rivets should be about $1.5t$ (or $1.5t'$) in diameter. Then

$$d = 1.5t' = 1.5 \times 0.563 = 0.844 \text{ in.}$$

Use $\frac{7}{8}$-in.-diam rivets.

The distance between gage lines (transverse pitch) should be

$$p_t = 2.5d = 2.5 \times 0.875 = 2.188 \text{ in.}$$

Make $2\frac{1}{4}$ in.

$$\text{Margin} = 1.5d = 1.5 \times 0.875 = 1.313 \text{ in.}$$

Make $1\frac{1}{2}$ in.

The sum of the two strap thicknesses should be greater than that of the main plate. Therefore,

$$\frac{0.563}{2} = 0.282 \text{ in.}$$

Make straps $\frac{5}{16}$ in. = 0.313 in.

3. Although the net area for any repeating section on the inside gage line is less than that on the outside gage line, a tension failure in the plate would occur on the outside line, because such failure on the inside line would also involve some type of failure (other than tension) on the outside line. Therefore, for tension,

$$P = [p_1 - (0.875 + 0.125)]0.563 \times 10,000$$

and for shear,

$$P = 0.785 \times 0.875^2 \times 2 \times 3 \times 8500$$

By equating these two values,

$$[p_1 - (0.875 + 0.125)]0.563 \times 10,000$$
$$= 0.785 \times 0.875^2 \times 2 \times 3 \times 8500$$

and solving for p_1,

$$p_1 = \frac{0.600 \times 2 \times 3 \times 8500}{0.563 \times 10,000} + 1.00 = 5.43 + 1.00 = 6.43 \text{ in.}$$

Make $6\frac{1}{2}$ in.

4. The actual safe loads are:
Tension in plate:

$$P = (6.50 - 1.00)0.563 \times 10,000 = 31,000 \text{ lb}$$

Rivet shear:

$$P = 0.785 \times 0.875^2 \times 2 \times 3 \times 8500 = 30,600 \text{ lb}$$

Bearing in plate:

$$P = 0.563 \times 0.875 \times 3 \times 20,000 = 29,600 \text{ lb}$$

5. The efficiency is

$$\eta = \frac{\text{strength of joint}}{\text{strength of solid plate}} = \frac{29,600}{0.563 \times 6.5 \times 10,000} = 80.9\% \quad (37)$$

This is slightly higher than the assumed efficiency of 80 per cent. No redesign is necessary. If the efficiency had come out below the assumed value, changes would have to be made to raise it, as, for example, by shortening the pitch distance or increasing the rivet diameter.

The various dimensions are as follows:

Wall thickness $t' = \frac{9}{16}$ in.
Rivet diameter $d = \frac{7}{8}$ in.
Transverse pitch $p_t = 2\frac{1}{4}$ in.

First pitch $p_1 = 6\frac{1}{2}$ in.
Second pitch $p_2 = 3\frac{1}{4}$ in.
Margin $= 1\frac{1}{2}$ in.
Strap thickness $= \frac{5}{16}$ in.

It will be remembered that a cylindrical vessel need be only half as thick to guard against a circumferential break as against a longitudinal break. The girth (circumferential) joints, therefore, need have no more than half the strength of the longitudinal joints. For this reason, the girth joints are frequently riveted *lap* joints which have lower efficiencies but are cheaper to make.

PROBLEMS

1. Calculate the minimum wall thickness for a boiler, 5 ft 0 in. in diameter and 10 ft long, to withstand a steam pressure of 250 psi. Assume a design stress of the material in tension of 12,000 psi.

2. A cylindrical pressure vessel, 4 ft 6 in. in diameter, has a wall thickness of 0.5 in. Investigate to determine whether this thickness is sufficient to hold safely a steam pressure of 200 psi. Use a design stress of 11,000 psi for tension.

3. Steel pipe, 10 ft 6 in. in diameter, is used for penstocks in a hydroelectric plant to guide the water from the intake at the top of the dam to the turbines, a vertical distance of 200 ft. Calculate the necessary wall thickness at the turbines. Assume a design stress for the pipe material in tension of 10,000 psi. Water weighs 62.4 lb per cu ft.

4. A number of 8 ft 0 in. bronze spherical pressure vessels are to be installed in a chemical plant. The pressure on the inside of each of the spheres will be 100 psi. Calculate the required shell thickness. The design stress for bronze is 7000 psi.

5. A boiler with a joint efficiency of 75 per cent has a diameter of 5 ft 0 in. and a wall thickness of $\frac{3}{4}$ in. Investigate to determine whether this is sufficient to withstand a pressure of 400 psi. The design stress is 10,000 psi.

6. Butt-welded joints are to be used for a boiler 6 ft 6 in. in diameter to hold a steam pressure of 300 psi. The boiler plate is $1\frac{1}{4}$ in. in thickness. What must be the efficiency of the weld? Use a design stress of 10,000 psi.

7. The fillet-welded lap joint shown in the accompanying figure must hold safely a tensile load of 18,000 lb. Calculate the required length of the welds along the 4-in.-wide strap. Assume that the weld can develop a safe shearing stress of 7000 psi. The strap is $\frac{3}{8}$ in. thick.

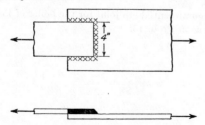

8. Two plates, $\frac{3}{4}$ by 10 in. in cross section, are to be butt-welded. The safe load for the joint is estimated as 82,000 lb. The design stress for the steel in tension is 11,500 psi. Calculate:

  a. The safe tensile load for the plates

  b. The efficiency of the joint

9. Two plates 12 in. wide are to be butt-welded and a tensile load of 90,000 lb is to be applied. Calculate the required thickness of the plates. Assume a design stress of the steel plates in tension of 12,000 psi and an efficiency of the weld of 85 per cent.

10. A riveted butt joint with three 1-in.-diam rivets joining each $\frac{1}{2}$-in. plate failed under test in plate bearing. The maximum load was recorded as 175,000 lb. Calculate the ultimate bearing strength of the material.

11. How many $\frac{7}{8}$-in.-diam rivets are needed in a lap joint with $\frac{5}{8}$-in. plates under a compressive load of 52,000 lb to guard against failure of the rivets in shear? Use a design stress of 8500 psi.

12. A lap joint has ten $\frac{3}{4}$-in.-diam rivets in two rows parallel with the edges of the plates. The cross-sectional dimensions of the plates are $\frac{1}{2}$ by 15 in. Calculate:

  a. The safe load for the joint

  b. The efficiency

Note: For this and other problems on riveted joints, use the design-stress values given in the text.

13. A butt joint with $\frac{7}{8}$-in.-diam rivets has eight rivets in two equal rows on each side of the butt. The main plates are $\frac{5}{8}$ in. thick and 12 in. wide; the straps are $\frac{3}{8}$ in. thick. Calculate:

  a. The safe load

  b. The efficiency

14. Two $\frac{3}{8}$- by 12-in. plates are to be fastened either by a butt joint using four $\frac{7}{8}$-in.-diam rivets on each side of the butt or by a lap joint using four 1-in.-diam rivets, whichever is stronger. Make calculations to determine which type should be used. Calculate the efficiencies.

15. The design of a riveted butt joint for a boiler calls for two rows of staggered rivets on each side of the butt. The pitch is the same for each row. Main plates are $\frac{3}{4}$ in. thick. Determine:

  a. The rivet diameter

  b. The margin

  c. The pitch (minimum)

  d. The transverse pitch

16. Design the thickness of boiler wall and a longitudinal butt joint with a rivet layout similar to that of Fig. 115. The boiler diameter is 5 ft 6 in. and the pressure is 225 psi.

17. Design a circumferential lap joint for the boiler of Prob. 16. Use two rows of rivets arranged as in Prob. 15.

CHAPTER 18

SCREWS, FASTENINGS, AND SEALS

95. Fastening Screws—Types and Materials. Bolts and screws are used to fasten parts together into a unit which must at times be disassembled. For permanent connection, it is advisable to fasten the parts by riveting, welding, or brazing.

In general, a bolt has two parts, the bolt and the nut, whereas a screw[1] is minus the nut. Figure 116 shows some of the numerous types of bolts and screws. The *through bolt* of Fig. 116a goes through both pieces that are held together. This type is made with square or hexagonal heads and nuts. A rough through bolt is a *machine bolt,*

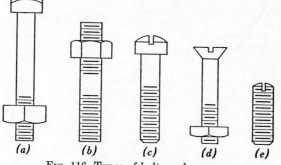

| (a) | (b) | (c) | (d) | (e) |

FIG. 116. Types of bolts and screws.

and one that is finished is a *coupling bolt.* A *stud bolt* (also called merely a *stud*) is shown in Fig. 116b. This bolt holds parts together where the hole can be drilled only through one of them. There is no head on the bolt and the thread extends for almost the entire length of the bolt shank. One end of the stud is screwed into one of the parts to be fastened, and the other end extends through the second part. The nut can then be tightened. A *cap screw* (Fig. 116c) is also used for this purpose. In addition to square and hexagonal heads, cap screws are made with countersunk, fillister, and socket heads.

[1] The term *screw* is also used in a general sense to mean both screws and bolts.

This type of screw is finished all over. *Machine screws* are similar to cap screws but are made in sizes up to ⅜-in. diam only. The heads are slotted. The *stove bolt* (Fig. 116*d*) *is* countersunk and slotted and is made in small sizes only. The *setscrew* of Fig. 116*e* is used to fasten the hub of a light pulley or some such part to a shaft to prevent independent rotation. It is screwed through a tapped hole in the hub of the pulley and tightened up on the shaft. Sometimes a recess is made in the shaft to accommodate the end of the screw, thereby

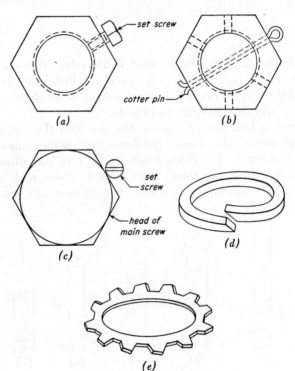

FIG. 117. Locking devices.

increasing the holding force. These screws also come in various types of heads.

Most screws are made of steel, usually plain carbon or nickel, with a carbon content ranging from low to medium. Sometimes screws are casehardened for increased resistance to wear. Special demands, as, for example, resistance to corrosion, call for screws of stainless steel or nonferrous metals, such as bronze.

96. Locking Devices and Washers. Various means are employed to prevent the nut or the screw from loosening up. We shall examine a few of the more common of these so-called locking devices. In Fig. 117a the setscrew may allow some slippage of the nut on the bolt; slippage may occur also in the lock washers in Fig. 117d and e. But the devices shown in Fig. 117b and c are positive and no slippage is possible. Lock washers have sharp edges or prongs to dig into metal

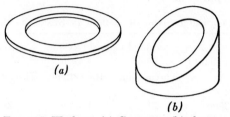

(a)

(b)

FIG. 118. Washers. (a) Common; (b) slant.

on each side. Furthermore, the spring action maintains a high frictional resistance to loosening of the nut even when some loosening has taken place.

Washers are flat rings used between the head of a screw and the metal to be held or between the nut and the metal. Washers serve several purposes. They may be used to prevent surface injury to parts or to serve as take-up where there is insufficient thread on the bolt shank. Special *slant washers* (Fig. 118b) prevent bending of the shank when the surface of the metal is not perpendicular to the hole.

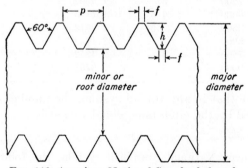

FIG. 119. American National Standard thread.

97. Screw Threads. The standard type of thread for bolts and screws in use in the United States is called the American National Standard and is illustrated in Fig. 119. It is important to know the following symbols and definitions pertaining to threads.

1. n is the number of threads per inch measured parallel to the long axis of the screw.

2. p is the *pitch* and is the distance between any point on one thread and the corresponding point on the next thread.

3. l is the lead which is the distance that the nut advances axially for each revolution. For single screw threads, the lead and pitch are the same.

4. h is the *depth of thread*.

TABLE 9. SELECTED LISTING OF AMERICAN STANDARD SCREW THREADS

Size	Major diam	National Coarse (NC)		National Fine (NF)	
		Threads/in.	Root diam	Threads/in.	Root diam
¼	0.2500	20	0.1850	28	0.2036
⁵⁄₁₆	0.3125	18	0.2403	24	0.2584
⅜	0.3750	16	0.2938	24	0.3209
⁷⁄₁₆	0.4375	14	0.3447	20	0.3725
½	0.5000	13	0.4001	20	0.4350
⁹⁄₁₆	0.5625	12	0.4542	18	0.4903
⅝	0.6250	11	0.5069	18	0.5528
¾	0.7500	10	0.6201	16	0.6688
⅞	0.8750	9	0.7307	14	0.7822
1	1.0000	8	0.8376	14	0.9072
1⅛	1.1250	7	0.9394	12	1.0167
1¼	1.2500	7	1.0644	12	1.1417
1⅜	1.3750	6	1.1585	12	1.2667
1½	1.5000	6	1.2835	12	1.3917
1¾	1.7500	5	1.4902		
2	2.0000	4.5	1.7113		
2¼	2.2500	4.5	1.9613		
2½	2.5000	4	2.1753		
2¾	2.7500	4	2.4252		
3	3.0000	4	2.6752		

You can see that in any thread system, the number of threads per inch multiplied by the pitch must equal 1, that is,

$$np = 1 \qquad (57)$$

In the American National Standard system, the following relations hold:

$$h = 0.6495p \qquad (58)$$

and

$$f = \frac{p}{8} \qquad (59)$$

where f is the width of the flat portion on the crest and trough of each thread. Among other advantages, the removal of these sharp angles lessens stress concentration.

One group of American Standard threads is the National Coarse series (NC) and another is the National Fine series (NF). To designate a thread, the major diameter, number of threads per inch, and series are given, for example, 1/2-13 NC. As can be noted from Table 9 for any major diameter, there are more threads per inch in the NF column. This also means that the NF minor diameters are larger. In addition to the above two series there is an extra-fine series called the SAE Extra Fine. These have still smaller threads and more threads to the inch. More threads to the inch means increased friction and less chance of jarring loose. Larger root diameter means larger area of cross section and a stronger screw for stresses on this area. In general, NC threads are used for nonferrous metals and cast iron, whereas NF threads are used for steel to steel. The larger threads of the NC series mean that each thread is stronger. This fact recommends the use of NC threads for brittle materials as a safeguard against failure from shock loads, as in the case of cast iron. The

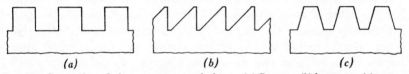

| *(a)* | *(b)* | *(c)* |

Fig. 120. Screw threads for power transmission. (*a*) Square; (*b*) buttress; (*c*) acme.

tapped holes of the cast-iron cylinder block of a gas engine have NC threads. Job requirements in large part dictate the thread series to use.

In this connection the fit of the nut or tapped hole with the threaded shank should also be mentioned. There are four standard classes of fits ranging from loose to very close[1]. Other considerations being equal, the tighter the fit, the greater is the frictional resistance against loosening. Again the job requirement is the controlling factor.

98. Screws for Power Transmission. Thus far screws for fastening have been discussed exclusively. Another type of screw is that used for the transmission of power, such as the screws of screw jacks, vises, and testing machines. The worm is a transmission screw with its thread in the form of a gear tooth. A number of the problems on force and motion presented in Chap. 5 deal with transmission screws.

The triangular cross sections of the American National Standard series produce a component thrust parallel to the side of the triangle and in an outward direction when a turning moment is applied. This thrust tends to burst the tapped hole. Hence, this type of thread is

[1] Thus a 1/2-13 NC thread with a class 2 fit would be shown as 1/2-13 NC-2.

not suitable for power transmission and other types are used, as shown in Fig. 120.　The *square thread* and the *buttress thread* are the best. The buttress thread is used for pushing on the perpendicular side of the thread only.　Although not as suitable as the others, the *acme thread* is cheaper to make.

99. Stresses in Bolts and Screws.　Stresses in bolts and screws are of two origins:

1. Stresses induced by tightening the nut or screw
2. Stresses induced by outside forces on the parts held together

Tightening of the nut or screw induces stresses both of tension and torsion.　It can easily be seen that tightening a screw stretches it and induces tension.　Furthermore, because of friction, the screw resists being tightened by the wrench or screwdriver to create a torsional stress.　Many of us have had the experience of shearing the cross section of a wood screw while driving it into a highly resistant piece of wood.

Various tests have been made and formulas developed for calculating the value of these stresses.　However, none of these formulas are satisfactory because the amount of tightening varies with the tightener.　Good practice is to keep these stresses at a minimum by applying no more force on the wrench than needed to tighten up[1] and by using some locking device to prevent loosening.

The induced stresses from outside forces are mainly tension or shear, or more usually a combination of both.　In designing for the proper area of shank cross section to withstand such stress, we must remember that the area calculated must correspond to the root area of the screw. From this the root diameter is obtained.　For simple stresses of tension or shear, we may apply the equations

$$A = \frac{P}{s} \tag{1}$$

and

$$d = \sqrt{\frac{A}{0.785}}$$

The next step is to refer to the table of screw threads and select the most suitable standard thread and size of screw.

When a stress of tension exists in the shank of the screw, the threads of the screw and the tapped hole or nut are stressed in shear.　Failure has been known to occur when there is insufficient area resisting this

---

[1] By the use of a torque wrench for tightening, the degree of tightness can be measured.

shearing stress, that is, when an insufficient number of threads are engaged. However, for the full engagement of a nut or corresponding engagement in a tapped hole, the shearing strength of the threads so far surpasses the tensile strength of the shank that there is no possibility of a shearing failure of the threads.

100. Factor of Safety and Design Stress. We have already mentioned the difficulty of determining the stress value caused by tightening the screw. Although no formulas for calculating this value are herein presented, the effect cannot be disregarded. Hence, an increased factor of safety is called for. Stress concentration also must be taken into account, by either a stress-concentration factor or an increased factor of safety. The crosses on Fig. 121 mark points of

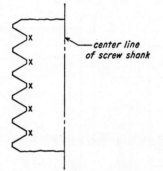

FIG. 121. Stress concentration in screw thread.

greatest stress concentration. Another consideration is the possibility of impact loads and repeated stress.

All these considerations add up to a very large factor of safety and a very low design stress. Values of design stress should range from around 2000 psi for screws up to $\frac{1}{2}$ in. in diameter to 4000 psi for screws up to 1 in. in diameter and over. With smaller-diameter screws there is more likelihood that the stress caused by tightening up is greater. If a limit stress of 30,000 psi is assumed, then a design stress of 3000 psi brings the factor of safety to 10.

Illustrative Example. Two screws for a pipe hanger must hold a tensile load of 1800 lb. Calculate the most suitable size. Use the NC series.

$$\text{Load on each screw} = {}^{180}\!\%_2 = 900 \text{ lb}$$

From inspection we can estimate that screws between $\frac{1}{2}$ in. and 1 **in.** diam will be needed. Choose a design stress of 3000 psi.

$$A = \frac{P}{s} = \frac{900}{3000} = 0.300 \text{ sq in.} \qquad (1)$$

$$d = \sqrt{\frac{A}{0.785}} = \sqrt{\frac{0.300}{0.785}} = 0.617 \text{ in.}$$

From Table 9 we select a 3/4-10 NC with a root diameter of 0.6201 in.

101. Screw Brackets. The bracket of Fig. 122 is to be fastened to the flange of a steel column by means of four bolts as shown.[1] The problem is to determine the proper size screws to use.

Point A is the fulcrum point, and, to maintain equilibrium, the moment of the 1000-lb force must be balanced by the resisting moments of the bolts. Any load on the bracket causes some stretching of the bolts with the top two stretching more than the bottom two. Since

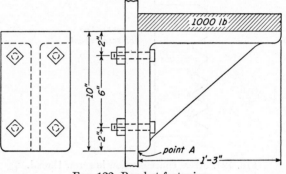

FIG. 122. Bracket fastening.

the frame is assumed to be rigid, the amount of stretch in each case is in proportion to the distance from point A. But we know that stress is proportional to strain up to the elastic limit, and therefore the stresses as well as the forces on the bolts are in proportion to the distances from point A. The value of each force can then be figured in terms of the others and the distances. Let the force on the lower set of bolts be designated by F_1 and that on the upper set by F_2. Then

$$\frac{F_1}{F_2} = \frac{2}{8}$$

$$F_2 = 4F_1 \quad \text{and} \quad F_1 = \frac{F_2}{4}$$

Now we shall take moments about point A.

$$1000 \times 7.5 - 2F_1 - 8F_2 = 0$$

[1] This design analysis is applicable to both screws and rivets.

and, by substituting $4F_1$ for F_2,

$$1000 \times 7.5 - 2F_1 - 8 \times 4F_1 = 0$$
$$F_1 = {}^{7500}\!/_{34} = 221 \text{ lb}$$
$$F_2 = 4 \times 221 = 884 \text{ lb}$$

In addition to the tensile forces just determined, there is also a shearing force of 1000 lb divided equally among all the bolts. Since the tensile force and shearing force on each bolt act on planes normal to each other, the true maximum force that must be resisted is not the sum of the two, but the sum of the two is an approximation which is on the safe side.[1] Then

$$\text{Force on each lower bolt} = \frac{221 + 500}{2} = 361 \text{ lb}$$

$$\text{Force on each upper bolt} = \frac{884 + 500}{2} = 692 \text{ lb}$$

It is usual practice in this type of connection to use the same size bolts top and bottom. From inspection the required area seems to be somewhat over $\frac{1}{2}$ sq in. and we shall assume a design stress of 3000 psi.

$$A = \frac{P}{s} = \frac{692}{3000} = 0.231 \text{ sq in.} \tag{1}$$

$$d = \sqrt{\frac{A}{0.785}} = \sqrt{\frac{0.231}{0.785}} = 0.543 \text{ in.}$$

By reference to Table 9 we note that a 5/8-18 NF bolt with a root diameter of 0.5528 in. is the most suitable.

102. Gasket Joints and Ground Joints. Screws often serve to hold together parts of pressure vessels which, unlike boilers, must sometimes be taken apart (Fig. 123). The cylinder block of an automobile is an example in which the head of the block is attached to the block proper by means of screws. The gasket separating the two parts in *gasket joints* (Fig. 123a) is usually two thin metal sheets covering a thicker asbestos filler. The purpose of the gasket is to ensure a leakproof joint. Other materials used for lower pressures include rubber, paper, and compressed cork with a binder. In *ground joints* the gasket is eliminated and there is metal-to-metal contact (Fig. 123b). This type of joint is used for very high pressures. The metal surfaces must be very precisely machined to prevent leakage, and consequently ground joints are quite costly.

---

[1] There exists a maximum tensile stress and a maximum shearing stress. Formulas for obtaining their values are derived in more advanced texts on strength of materials.

When the bolt of the ground joint is tightened, as it must be to function properly, the shank has stretched slightly and acts as a very powerful tension spring in holding the two metal lips together. If pressure is then introduced within the vessel, there would also be an external tensile force applied to the bolt. However, the bolt will stretch no further as long as the external force is below that of the initial tightening force on the bolt. This means that there is no additional tensile stress induced, because there can be no additional stress without additional strain.

Bolts holding gasket joints of elastic materials act differently. As the nut is screwed tight, the bolt stretches a little, but mostly the elastic gasket material is pushed together and shortened and acts as a spring on the bolt. When pressure is introduced inside the vessel, the result this time is a stretching of the bolt almost in proportion to the

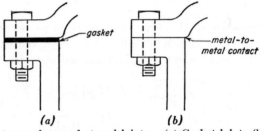

(a)   (b)

Fig. 123. Two types of screw-fastened joints. (a) Gasket joint; (b) ground joint.

tensile load produced by the pressure. The elongation of the bolt is taken up by the expansion of the gasket. Stretching of the bolt (strain) means that more stress is induced therein.

The above analyses would lead us to believe that more stress is induced on screws of joints where an elastic gasket is used than where there is metal-to-metal contact. In fact, some designers make such allowances. However, we must also consider the fact that the elastic gasket bolt is not screwed up as tight as the ground-joint bolt, and therefore less initial stress is induced. One factor may be regarded as roughly balancing the other, and it is here submitted as inadvisable in design to consider the elastic gasket bolt as resisting more load than the ground-joint bolt. As previously stated, it is believed that the low design stress recommended is sufficient to allow for the induced stress of tightening up. It is, therefore, suggested that bolts and screws of pressure vessels should normally be considered to resist the external forces only.

103. Stuffing Boxes. In reciprocating steam engines and pumps, the piston rod must go through a hole in the wall of the cylinder to the óutside and, while the engine is in operation, the rod moves back and forth through this hole. The enclosing fluid will escape between the hole and the rod unless properly sealed. This sealing or packing material is held in place by a *stuffing box*. Figure 124 illustrates one type of stuffing box. The packing material is placed around the piston rod in the space provided and the *gland* is screwed into the boss on the

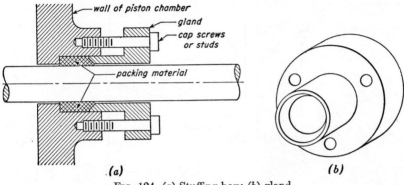

.(a)   (b)

Fig. 124. (a) Stuffing box; (b) gland.

piston chamber. Too much tightening of the screws means excessive friction and too little invites leakage.

The type of packing material selected varies with the nature of the fluid under pressure in the piston chamber. Asbestos rope which has been graphitized is used for steam and hot water. Cotton rope soaked in oil is satisfactory for cold fluids.

PROBLEMS

1. Calculate the angle (helical angle) of a thread on a square-threaded screw, 2 in. in diameter. The screw is triple-threaded with a lead of $\frac{3}{4}$ in.

2. Calculate the lead, pitch, width of crest, depth of thread, and minor diameter for a double-threaded 1-8 NC screw.

3. Plates of a lap joint are fastened together by cap screws with fully threaded shanks. The joint must hold a tensile load of 25,000 lb. Calculate the number of 3/4-16 NF cap screws needed to ensure against a shearing failure of the screws. Use a design stress for shear in the screws of 2500 psi.

4. A bearing bracket is fastened to the lower flange of an overhead steel beam with four 7/8-14 NF bolts. The shaft transmits a tensile load of 7500 lb to the bolts. Make calculations to determine whether the bolts can hold this load with safety. Assume a design stress of 3500 psi for the bolts.

5. Six studs are used to resist a tensile force of 8500 lb. Calculate the proper size bolts for an NC thread. Assume a design stress of 3000 psi.

6. A 2-in.-diam alloy-steel piston rod extends through the piston of a steam engine and holds the piston on one side by a shoulder of the rod and on the other by a locked nut. The maximum steam pressure developed is 200 psi and the inside diameter of the piston chamber is 8 in. The limit stress of the steel is 40,000 psi with a factor of safety of 6 specified.

a. Calculate the necessary root diameter for the rod.

b. Select the proper NF thread for this root diameter.

7. Calculate the number and size of bolts for the three-hole bracket shown in the accompanying figure. Use an NC thread and a design stress of 3500 psi.

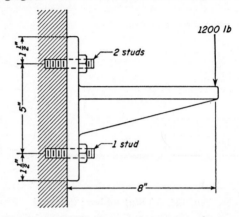

8. A plate held by four 1-8 NC screws seals a rectangular opening 6 in. by 5 in. in the steam chest of a pump. What is the maximum pressure permissible within the chest? Assume a design stress for the bolts of 4000 psi.

9. The cylinder of a hydraulic press has a diameter of 10 in. The water pressure is 150 psi. There are to be eight standard NC screws to hold the cylinder head. Calculate the proper screw size, assuming a design stress of 2500 psi.

CHAPTER 19

SPRINGS

104. Uses, Types, and Materials. The uses of springs are many and varied. In trains, automobiles, and trailers, springs are used to absorb shock and vibration. In clutches, springs help to transmit power by increasing frictional resistance. In doors, springs cause motion, whereas the spring action of a lock washer prevents motion. In spring balances, in some testing machines, and in steam-engine indicators, springs measure the magnitude of the forces. In clocks, watches, and a variety of recording devices, springs furnish the motive power.

Types vary with the varied uses. Figure 125 shows some of the more common types of springs. The flat spring here illustrated acts as a cantilever beam and the leaf spring as a simply supported beam.

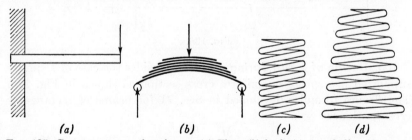

(a)	(b)	(c)	(d)

Fig. 125. Common types of springs. (a) Flat; (b) leaf; (c) true helix; (d) cone-shaped helix.

The helical type of coil springs are used to resist tensile, compressive, or twisting forces. Those shown are compression springs.

Spring steel is often plain carbon steel with a carbon content varying from medium to high (about 0.50 to 1.00 per cent). It is specially heat-treated. The carbon content for leaf springs runs somewhat higher than for coil springs. Special alloy steels with outstanding properties of resilience are also used. These include chrome-vanadium and silicon-manganese steels, with the latter mostly for leaf springs. Phosphor-bronze springs are very resilient and resist corrosion.

105. Spring Loadings and Design Stresses. The factors of dynamic loading, repeated loading, and stress concentration which have been discussed in connection with the design of various machine parts are of great importance in the design of springs. Many springs, similar to springs on a friction clutch, are continuously subjected to load application and removal. Springs attached to automobile axles, for example, in addition to the application of repeated loads, must withstand the shocks caused by bumps and depressions in the roadbed. In coil springs, stress concentration should be taken into account. The sharper the bend of the coil, the more important this factor.

As in the case of other machine parts, these factors should be reflected in a larger factor of safety and lowered design stress based on the elastic or the endurance limit as conditions dictate. Despite such precautions springs do break, as every automobile owner knows. It is difficult to present values for this design stress. In chrome-vanadium steels with an elastic limit as high as 90,000 psi, the design-stress value may be 30,000 or 40,000 psi. On the other hand, for the plain carbon steels the value may be 10,000 to 20,000 psi.

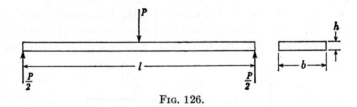

FIG. 126.

106. Design of a Flat Spring. Let us consider the case of a simply supported flat spring of uniform cross section as shown in Fig. 126. The flexure formula, developed in Sec. 71 for beams of rectangular cross section, is

$$\frac{M}{s} = \frac{bh^2}{6} \tag{45}$$

From inspection we note that the point of zero shear (point of maximum bending moment) is at the mid-section. Therefore,

$$M = \frac{P}{2} \times \frac{l}{2} = \frac{Pl}{4}$$

and, by substitution,

$$\frac{Pl}{4s} = \frac{bh^2}{6}$$

By solving for h and reducing,

$$h = \sqrt{\frac{3Pl}{2bs}} \tag{60}$$

In this form of the equation, the depth of the spring can be determined when the other terms are known.

Illustrative Example. Calculate the maximum safe load that can be applied to the mid-section of a flat spring supported as a simple beam. The material is plain carbon steel and the spring is of uniform cross section with a length of 18 in., a width of 3 in., and a thickness of $\frac{5}{8}$ in. Use a design stress of 15,000 psi.

$$h = \sqrt{\frac{3Pl}{2bs}} \tag{60}$$

$$P = \frac{2h^2bs}{3l} = \frac{2 \times 0.625 \times 0.625 \times 3 \times 15,000}{3 \times 18} = 650 \text{ lb}$$

Figure 127a is an exaggerated representation of the bending in a flat cantilever beam spring of uniform cross section. Each cross section has the same resisting strength in bending, but the bending moment and, hence, the induced bending stress decreases in proportion

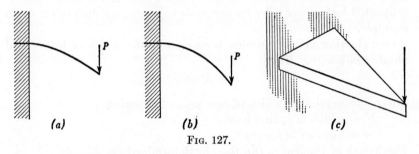

(a)   (b)   (c)

Fig. 127.

to the distance from the support. Moreover, the deflection curve flattens out to a straight line.

Note the deflection curve of Fig. 127b which is an arc of a circle (or very nearly so). A small portion of the length at the free end deflects as much as one at the support end. This spring is more "springy" (resilient), or, to put it another way, can store more energy than the spring of Fig. 127a. This is a desirable characteristic and in order to obtain it, each section of the spring must be stressed to an equal amount, which means that the *strength* in bending of the beam must decrease proportionately to the distance from the support to match the decreasing bending moment. The shape shown in Fig. 127c is the answer. Here the depth remains the same but the width varies. A glance at the flexure formula will verify the fact that bending strength is in proportion to the width b of the beam.

107. Leaf Springs. Another way of varying the width is to fasten springs of varying length on top of one another to form a laminated, or leaf, spring. The proper length for the leaves can be seen from a study of Fig. 128a and b. In each case the top view is shown. You can see that the leaf springs can be thought of as varying width beams, cut into strips, and the strips piled on top of each other. Since the simply

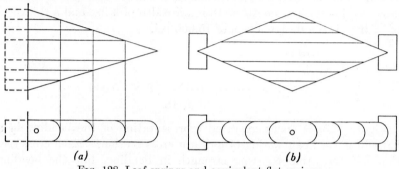

(a)   (b)

FIG. 128. Leaf springs and equivalent flat springs.

supported leaf spring is by far the most common, we shall use it as an example.

One starting point in design is to assume the width of leaves and number of leaves. Then

$$b = b' \times n$$

where b = maximum width of equivalent flat spring
b' = width of spring leaves
n = number of leaves
The length of the leaves can then be determined graphically.

Illustrative Example. An alloy-steel leaf spring for an automobile must sustain a force of 1000 lb at its mid-section. The supports are 30 in. apart. Calculate the spring dimensions, assuming a design stress of 30,000 psi.

Let us assume that the leaf width is $2\frac{1}{2}$ in. and that there are six leaves. This is a fair assumption for an automobile.

$$b = b' \times n = 2.5 \times 6 = 15 \text{ in.}$$

Next we draw the equivalent flat spring to an appropriate scale (Fig. 129) and divide it into strips. The middle strip is $2\frac{1}{2}$ in. wide, but the other strips are half as wide because each is laid out on both

sides of the middle strip.   Then the leaf spring is laid out below to the
same scale, the leaves are drawn in, and the ends rounded off.

$$h = \sqrt{\frac{3Pl}{2bs}} = \sqrt{\frac{3 \times 1000 \times 30}{2 \times 15 \times 30,000}} = 0.316 \text{ in.} \quad (60)$$

Use ⅜ in. depth.

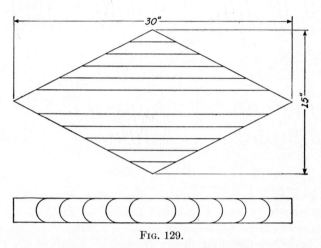

Fɪɢ. 129.

108. Leaf-spring Construction.   The greatest shear in a simple
beam occurs at the reactions.   In the leaf spring, the main, or master,
leaf must withstand this shearing force (refer to Fig. 130).   In addition there are twisting forces at the reactions because of initial non-
alignment when the pins are driven through the eyes, or because of

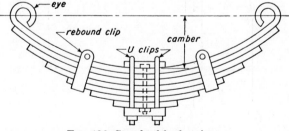

Fɪɢ. 130. Standard leaf spring.

sidewise thrusts.   To take these induced stresses into account in
design, this leaf is sometimes made of greater depth or the next leaf
made full length.   A bolt through holes drilled at the mid-section of
each leaf fastens the leaves together as do also the U clips.   The weak-

ening effect of the hole is lessened by the action of the U clips. At the quarter points approximately, the rebound clips are fastened to help distribute more stress to the shorter leaves during rebound.

109. Coil Springs. Figure 131 shows coil springs of the helical type, a for tension, b and c for compression, and d for torsion. The

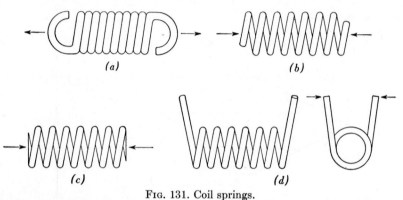

FIG. 131. Coil springs.

ground ends of c form a better bearing surface than the open ends of b. Strangely enough, a so-called tension, or compression, helical spring induces a stress of torsion principally, whereas the torsion spring induces both bending and direct tension or compression but not torsion.

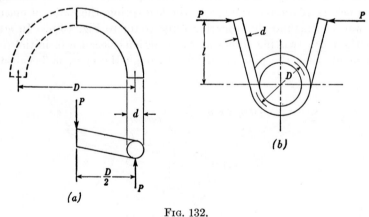

FIG. 132.

The torsional stress in either a tension or compression coil spring may be investigated by considering any one-quarter turn of the coil (Fig. 132a). The torque on this length is a maximum and must be resisted by the torsion in the cross section of the wire. We use the torsion formula for circular cross sections, namely,

$$s = \frac{16T}{\pi d^3} \qquad (12)$$

In our case

$$T = P \times \frac{D}{2}$$

Therefore, by substitution,

$$s = P \times \frac{D}{2} \times \frac{16}{\pi d^3}$$

or

$$s = \frac{8PD}{\pi d^3} \quad \text{and} \quad d = \sqrt[3]{\frac{8PD}{\pi s}} \qquad (61)$$

The symbols D and d should not be confused. They are the mean diameter of the coil and the diameter of the wire, respectively.

The torsion coil spring, shown in Fig. 132b, may be investigated as a beam in bending, considering the portion from the end (where the load P is applied) to the center line of the coil as a cantilever beam. The flexure formula applies.

$$s = \frac{Mc}{I} \qquad (23)$$

In this case

$$M = P \times l$$

From the table on page 215,

$$I = \frac{\pi d^4}{64}$$

also

$$c = \frac{d}{2}$$

By substitution,

$$s = P \times l \times \frac{d}{2} \times \frac{64}{\pi d^4}$$

which reduces to

$$s = \frac{32Pl}{\pi d^3} \quad \text{and} \quad d = \sqrt[3]{\frac{32Pl}{\pi s}} \qquad (62)$$

As previously discussed (Sec. 105), the value of the design stress s in Eqs. (61) and (62) depends to a certain extent upon the ratio of the diameter of the coil, D, and the diameter of the wire, d.

Illustrative Example 1. Calculate the diameter of wire needed in a coil spring to resist a tensile load of 70 lb. The coil diameter is set at 1 in. Use a design stress of 30,000 psi.

$$d = \sqrt[3]{\frac{8PD}{\pi s}} = \sqrt[3]{\frac{8 \times 70 \times 1}{3.14 \times 30,000}} = \sqrt[3]{0.00594} = 0.181 \text{ in.} \quad (61)$$

Use $\frac{3}{16}$ in. diam = 0.188 in.

Illustrative Example 2. Investigate a torsional spring with a wire diameter of $\frac{1}{4}$ in., to determine the safe load that it can support. This load acts $1\frac{1}{4}$ in. from the center line of the coil. The design stress is 35,000 psi.

$$s = \frac{32Pl}{\pi d^3} \quad (62)$$

$$P = \frac{\pi s d^3}{32l} = \frac{3.14 \times 35,000 \times 0.25 \times 0.25 \times 0.25}{32 \times 1.25}$$

$$= 43.0 \text{ lb}$$

110. Deflection in Springs. For flat or leaf springs, the maximum deflection for any known load can be calculated according to formulas based on the fundamental expression,

$$\text{Deflection } (\Delta) = \text{a constant} \times \frac{Pl^3}{EI}$$

For coil springs in tension or compression, formulas based on the fundamental expression for torsional deflection,

$$\alpha = \frac{57.3Tl}{GJ} \quad (26)$$

have been devised.

The *spring scale* is a means of comparing the deflection of one spring with another. It is defined as the force necessary to deflect a spring 1 in. To find the spring scale P_s, divide any force that is on a spring by the deflection that the force causes. In symbols,

$$P_s = \frac{P}{\Delta}$$

Forces which rest on springs (static loads) can easily be measured. However, if a weight is *dropped* on a spring (dynamic load), calculations must be made to determine the equivalent force. The maximum deflection can also be calculated. In the case of weights dropping on springs, the spring deflection cushions the effect of the fall, that is, the force is much less than if the weight were dropped on an unyielding

body[1].   Problems involving dynamic loads on springs occur frequently
in machine design.   For example, an automobile is traveling on a
bumpy road.   As a wheel hits a bump, that portion of the car above
the wheel is elevated, but immediately thereafter the car weight drops
on the spring.

Let us see how the spring deflection of a dynamic force can be calculated by means of the following example.

Illustrative Example.   A 100-lb weight is dropped from a height of
24 in. to a spring.   How much does the spring deflect?   The spring
scale is 100 lb.

Let Δ represent the deflection of the spring.   The total distance s
that the weight drops will then be

$$s = 24 + \Delta$$

The energy lost by the 100-lb weight is equal to the weight multiplied
by the distance through which it has moved, or

$$\text{Energy lost by weight} = 100(24 + \Delta)$$

Similarly, the work done on (or the energy gained by) the spring is the
average force P multiplied by the spring deflection, or

$$\text{Work on spring} = P \times \Delta$$

But the energy lost by the weight is equal to the energy gained by the
spring.[2]   Therefore

$$100(24 + \Delta) = P \times \Delta \tag{A}$$

The maximum force on the spring is twice the average, or $2P$.   It is
this force, attained at the end of the deflection, that caused the maximum deflection Δ.   Since a 100-lb force causes a deflection of 1 in.
(spring scale), we can set up a proportion.

$$\frac{2P}{\Delta} = \frac{100}{1}$$

and

$$2P = 100 \times \Delta$$

or

$$P = 50 \times \Delta \tag{B}$$

By substitution of the value of P of Eq. (B) in Eq. (A), we have

$$100(24 + \Delta) = 50 \times \Delta^2$$

[1] The reason for this is explained in Sec. 35.
[2] This is true if heat losses in the spring are neglected.

and

$$2(24 + \Delta) = \Delta^2$$

or

$$\Delta^2 - 2\Delta - 48 = 0$$

By factoring,

$$(\Delta - 8)(\Delta + 6) = 0$$
$$\Delta = 8 \text{ or } -6$$

We conclude that the spring will deflect 8 in.

PROBLEMS

1. Design a flat spring of uniform cross section to hold a load of 500 lb at its mid-section. The spring is supported as a simple beam. The width is limited to 3 in. and the length is 24 in. The material is plain carbon steel (AISI 1050). Use a design stress of 15,000 psi.

2. A flat cantilever spring of uniform cross section and with dimensions of $\frac{1}{2}$ by 2 by 10 in. supports a load of 400 lb at the free end. Make calculations to determine whether the induced stress is within the design-stress limit of 20,000 psi.

3. A leaf spring supported as a simple beam, 20 in. long, is required to hold a load of 800 lb. There are to be five leaves, each 2 in. wide.

a. Determine the leaf lengths by graphical means.

b. Calculate the leaf thickness. Use a design stress of 30,000 psi.

4. The two alloy-steel leaf springs on the rear axle of a truck have ten leaves each. The leaves are $\frac{5}{8}$ by 4 in. in cross section, and the distance center-to-center of supporting pins is 4 ft. Make calculations to determine whether the springs are sufficiently strong to support the rear-end load of 6 tons. The design stress is 25,000 psi.

5. A cantilever leaf spring is to be made with eight leaves, each $\frac{3}{8}$ in. thick. The spring must hold a load of 1200 lb at a distance of 16 in. from the support. A factor of safety of 2.5 based on an endurance limit of 70,000 psi is specified.

a. Calculate the required width of leaf.

b. Determine the leaf lengths.

6. Springs used as rebound snubbers in a railway car are $\frac{1}{2}$-in.-diam rods shaped into a helical coil. The mean diameter of the coil is 4 in. The load that each spring must support is 600 lb. Calculate the stress developed.

7. A coil spring is required to hold a tensile force of 40 lb. The ratio of D to d is set at 8:1, and the design stress is 32,000 psi. Calculate the wire diameter.

8. A load of 20 lb acts at a distance of $\frac{1}{2}$ in. from the center of a torsion spring. Calculate the required diameter of the spring wire. The design stress is 25,000 psi.

9. Calculate the safe load that can be applied to a torsion coil spring with a mean diameter of $\frac{3}{4}$ in. and a wire diameter of $\frac{3}{16}$ in. The center of gravity of the load is to be applied at a distance from the center of the coil equal to the mean coil diameter. Assume a design stress of 30,000 psi.

10. A 60-lb weight is dropped a distance of 9 in. to a spring. The spring deflects 3 in. Calculate the scale of the spring and the maximum force.

CHAPTER 20

COMBINED STRESSES

111. Combined Stresses in Machine Design. We have already discussed some cases of combined stresses, as, for example, in shafts and in screws, where torsion and bending, or torsion and tension, are induced simultaneously. The combined stresses were considered in the design. Also, as we remember from strength of materials, the loads on columns cause bending as well as direct compression. In order to take both into account, special column formulas are used. The machine designer often encounters other instances of more than one type of stress in a part for given loading conditions. A few of the simpler problems of combined stresses will be discussed in this chapter.

112. Tension and Bending—Symmetrical Sections. The first problem is the investigation of the crane hook of Fig. 133a. The circular cross sections AA, BB, and CC, as shown in the revolved sections, are all of equal area. At section AA the stress is tension only, at CC it is shear only, but at BB it is tension similar to AA and in addition bending stress, owing to the fact that BB is not on the line of action of the forces P. The total stress value at AA and CC is equal to the load P, whereas at BB the total stress is equal to P plus the bending stress. Since section BB is farthest from the force line of any cross section of the hook, it is the dangerous section and should be investigated.

Figure 133b represents section BB with the stress of *direct tension* only considered. This stress is assumed to be evenly distributed. The arrows indicate tension, that is, a pull on the section. Figure 133c represents the stress of *bending* only on the section, made up (as is true in all bending) of compression and tension, each starting at the neutral axis and increasing in proportion to the distance from that axis. The bending in this case causes tension on the inside of the hook and compression on the outside. The direct tension and the bending stresses are combined in Fig. 133d by adding b and c algebraically to show the true stress on section BB. Note that the stress is mostly tension with a smaller amount of compression at the right of the figure and that the maximum stress occurs at the extreme left.

At section BB let the stress of direct tension equal s_1, the maximum stress of bending equal s_2, and the maximum resultant stress equal s.

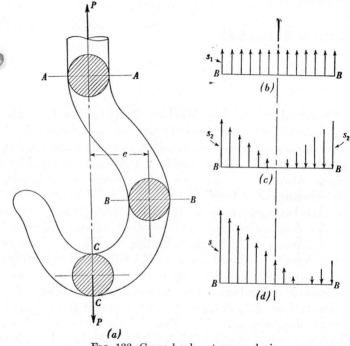

(a)

FIG. 133. Crane hook—stress analysis.

Then

$$s = s_1 + s_2$$

but

$$s_1 = \frac{P}{A} \tag{1}$$

and

$$s_2 = \frac{Mc^*}{I} \tag{23}$$

Hence,

$$s = \frac{P}{A} + \frac{Mc}{I}$$

* It must here be mentioned that the flexure formula, $M/s = I/c$, is strictly applicable to straight beams only and not to curved bending members. This is because the neutral axis shifts from the gravity axis in a curved bending member when under load. The formula for "curved beams" is quite involved and the result little different from that obtained by the use of the flexure formula.

The bending moment causing the stress of bending at section BB is the product of the load and the distance from the line of application of the load to the center of moments, considered to be at the neutral axis, that is, $M = Pe$. By substitution,

$$s = \frac{P}{A} + \frac{Pec}{I} \qquad (63)$$

Also

$$sAI = PI + PAec$$
$$sAI = P(I + Aec)$$

and

$$P = \frac{sAI}{I + Aec} \qquad (63)$$

Since s represents the greatest stress anywhere in the hook, s for safety must not be more than the design stress of the material. By substituting the design stress for s in the second form of Eq. (63), the safe load, P, can be determined.

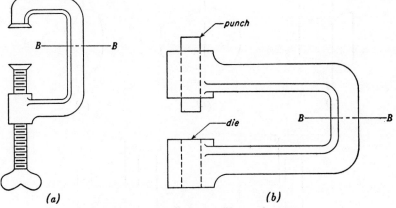

FIG. 134. (a) C clamp; (b) punch press.

Although the above analysis is built around a specific problem of a crane hook, there are other practical examples of the combined stresses of tension and bending for which the analysis is applicable. Figure 134a and b represent a C clamp and a punch press, respectively. In each case section BB is the dangerous section, subjected to both tensile and bending forces.

Machine shops and assembling plants are busy places, and crane hooks and such, as so many of the other machine parts previously discussed, are subjected to repeated load applications as well as to dynamic loads. We take these factors into account by a lowered design stress.

Illustrative Example. Calculate the maximum load that can be carried safely by the crane hook of Fig. 133a. Assume a design stress of 8000 psi. The hook is circular in cross section with a diameter of 1½ in. at the dangerous section. Maximum distance from the center line of the hook to the line of action of the forces is 2 in.

$$P = \frac{sAI}{I + Aec} \qquad (63)$$

From the table on page 215,

$$I = \frac{\pi d^4}{64} = \frac{3.14 \times 1.5 \times 1.5 \times 1.5 \times 1.5}{64} = 0.248 \text{ in.}^4$$
$$A = 0.785d^2 = 0.785 \times 1.5 \times 1.5 = 1.77 \text{ sq in.}$$
$$P = \frac{8000 \times 1.77 \times 0.248}{0.248 + 1.77 \times 2 \times 0.75} = \frac{3510}{2.90} = 1210 \text{ lb}$$

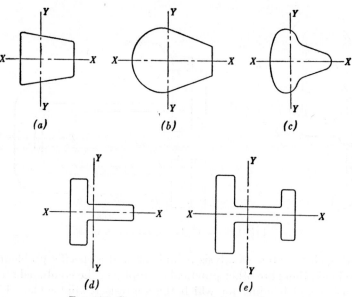

(a)   (b)   (c)

(d)   (e)

Fig. 135. Common unsymmetrical sections.

113. Tension and Bending—Unsymmetrical Sections. Crane hooks, C clamps, or punch presses can be strengthened by adding more material to the side of greatest stress. This makes the cross section of the dangerous section unsymmetrical about its neutral axis, similar to the shapes shown in Fig. 135. However, the investigation of the strength of unsymmetrical cross sections is more involved. In the first place, the moment of inertia about the center-of-gravity axis (neutral axis of the section) cannot be calculated until the center-of-gravity axis is located. All the sections illustrated in Fig. 135 are unsymmetrical about the yy axis. The problem is first of all to locate yy. A brief review of this topic, studied in strength of materials, is in order and will be made with reference to the shape shown in Fig. 135b (reproduced in Fig. 136) as a specific example.

First, it is to be noted that the section can be divided into a semicircle and a trapezoid, the areas of which we will denote as S and T, respectively. Then a reference axis, as the axis RF, is drawn parallel to yy. It is possible to place this parallel axis anywhere either inside or outside the section, but an axis through the extremity leads to a simple solution. The entire section is thought of as free to revolve about this axis. The *area moment* of each of the parts about the reference axis is the product of the area of that part and the distance from its center-of-gravity axis to the reference axis. In this case the area moments of the two parts are $S \times d$ and $T \times h$. Now consider the fact that the sum of the two area moments must be

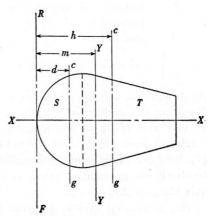

Fig. 136. An unsymmetrical section divided into two geometric figures.

equal to the area moment of the entire section, composed of the product of the area of the entire section and the distance from *its* center-of-gravity axis (yy) to the reference axis. In symbols,

$$(S + T)m = S \times d + T \times h$$

and

$$m = \frac{Sd + Th}{S + T}$$

When the distance m is calculated, the location of the yy axis is known. Putting the above expression in terms of a general formula applicable to all cases, we have

$$x_G = \frac{\Sigma \text{ area moments}}{\Sigma \text{ areas}} \tag{64}$$

where x_G represents the distance from the reference axis to the center of gravity of the compound section.

The moment of inertia must now be calculated. No formula for it can be found in a table, but in the table on page 215 we do find formulas for the semicircle and the trapezoid of moments of inertia about their own center-of-gravity axes. With these values known, a special formula,

$$I_{xx} = I_{cg} + Ay_I^2 \quad \text{or} \quad I_{yy} = I_{cg} + Ax_I^2 \tag{65}$$

is applied to each part (semicircle and trapezoid in this case). The meanings of the terms are as follows:

$\left.\begin{array}{c} I_{xx} \\ I_{yy} \end{array}\right\}$ = moment of inertia of part about new axis (xx or yy)

I_{cg} = moment of inertia of part about its center-of-gravity axis parallel to axis xx or yy

A = area of part

$\left.\begin{array}{c} y_I \\ x_I \end{array}\right\}$ = distance between the two parallel axes

As in algebra, vertical distances are y and horizontal distances are x. Knowing the moment of inertia of each part about its own center-of-gravity axis (I_{cg}), we can use the formula to obtain the moment of inertia of each part about the neutral axis of the compound section (I_{xx} or I_{yy}). This being accomplished, the new values are added up to obtain the moment of inertia of the entire section about the neutral axis (ΣI_{xx} or ΣI_{yy}).

The next step in investigation is to calculate the induced stress or the safe load. For the symmetrical sections there is no doubt that the maximum stress occurs at that point on the dangerous section where the greatest tensile stress of bending is developed, for example, on the inside of the crane hook. If the load is known, the maximum induced stress can be found according to the equation

$$s = \frac{P}{A} + \frac{Pec}{I} \tag{63}$$

The stress on the unsymmetrical sections follows a similar pattern but sometimes gives a different result.

Figure 137 shows a section unsymmetrical about the yy axis and an analysis of the stresses on it. Here the maximum compressive stress caused by bending (Fig. 137c) is greater than the tensile stress so caused. That this must be true is evident from the fact that the tension and compression of bending increase proportionately as the distance from the neutral axis, and that this axis of the section is nearer to the left side. In this case also the direct tension is small in compar-

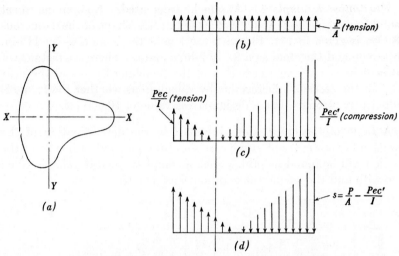

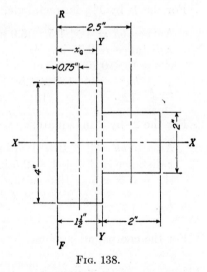

Fig. 137. Unsymmetrical section—stress analysis.

ison (because of a large eccentricity) and even though the direct tension is *subtracted* from the maximum compression of bending, and *added* to the maximum tension of bending, nevertheless, the resulting maximum stress is one of compression (Fig. 137*d*).

Therefore, in the investigation of unsymmetrical sections, another equation similar to (63) should also be used.   This is

$$s = \frac{P}{A} - \frac{Pec'}{I} \quad (63a)$$

and

$$P = \frac{sAI}{I - Aec'}$$

where c' is the distance from the neutral axis to the extreme fibers farther away.   Remember that it is not necessarily true that the com-

Fig. 138.

bination of the compressive stress caused by bending and the direct tension will be the maximum stress, but such *may* be the case, and therefore both possibilities should be investigated.   In other words, for unsymmetrical sections use *both* equations, (63) and (63*a*), to determine which gives the greater stress or the smaller load.

Illustrative Example 1. A punch press used for stamping sheet metal has a punching capacity of 5000 lb. The shape of the dangerous section is given in Fig. 138 with the top of the T as a 4- by 1½-in. rectangle and the stem as a 2- by 2-in. square. There is a distance of 10 in. between the line of action of the punching force and the near edge of the section. Determine by calculations whether the frame is sufficiently strong. Use a design-stress value of 10,000 psi.

A table of the calculations involved in the determination of the location of the center-of-gravity axis and the moment of inertia of the section will be found helpful. Such a complete filled-in table is given herewith and the calculations are outlined below.

Section	Area	x	Ax	I_{cg}	x_I	x_I^2	Ax_I^2	$I_{cg} + Ax_I^2$
▮	6.0	0.75	4.5	1.12	0.70	0.49	2.94	4.06
▪	4.0	2.50	10.0	1.33	1.05	1.10	4.40	5.73
Total	10.0	....	14.5	....	....	....	....	9.79

Assume the reference axis RF.

For the 4- by 1½-in. rectangle,

     Area $= bh = 4 \times 1.5 = 6.0$ sq in.
     x distance $= 0.75$ in.
     $Ax = 6 \times 0.75 = 4.5$ cu in.
$$I_{cg} = \frac{bh^3}{12} = \frac{4 \times 1.5 \times 1.5 \times 1.5}{12} = 1.12 \text{ in.}^4$$

For the 2- by 2-in. square,

     Area $= bh = 2 \times 2 = 4$ sq in.
     x distance $= 1.5 + 1 = 2.5$ in.
     $Ax = 4 \times 2.5 = 10.0$ cu in.
$$I_{cg} = \frac{bh^3}{12} = \frac{2 \times 2 \times 2 \times 2}{12} = 1.33 \text{ in.}^4$$

For the compound section,

$$x_G = \frac{\Sigma \text{ area moments}}{\Sigma \text{ areas}} = \frac{14.5}{10} = 1.45 \text{ in.} \qquad (65)$$

x_I (for 4- by 1½-in. rectangle) $= 1.45 - 0.75 = 0.70$ in.
x_I (for 2- by 2-in. square) $= 2.50 - 1.45 = 1.05$ in.
Distance e from c.g. of section to line of action of forces
     $= 1.45 + 10.00 = 11.45$ in.

Calculation of maximum stress:

For tension side,

$$s = \frac{P}{A} + \frac{Pec}{I} = \frac{5000}{10} + \frac{5000 \times 11.45 \times 1.45}{9.79}$$
$$= 500 + 8490 = 8990 \text{ psi} \tag{63}$$

For compression side,

$$s = \frac{P}{A} - \frac{Pec'}{I} = \frac{5000}{10} - \frac{5000 \times 11.45 \times 2.05}{9.79}$$
$$= 500 - 12{,}000 = -11{,}500 \text{ psi} \tag{63a}$$

Since the design stress is 10,000 psi and the maximum induced stress in the frame is 11,500 psi, the frame is of insufficient strength.

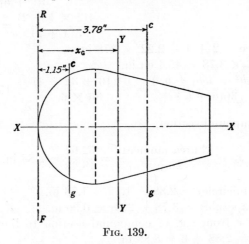

FIG. 139.

Illustrative Example 2.   Calculate the maximum load that can be placed with safety on a crane hook with the dangerous section as illustrated in Fig. 136 (reproduced in Fig. 139).   Assume design stress of 8000 psi.   The distance from the nearest point of this section to the line of action of the forces is 6 in.   The radius of the semicircle of the cross section is 2 in. and the height of the trapezoid is 4 in., with bases of 4 in. and 2 in.

Section	Area	x	Ax	I_{cg}	x_I	$x_I{}^2$	$Ax_I{}^2$	$I_{cg} + Ax_I{}^2$
◀	6.28	1.15	7.22	1.76	1.73	2.99	18.8	20.56
▶	12.00	3.78	45.4	15.4	0.90	0.81	9.72	25.12
Total	18.28	....	52.62	.....	....	....	.....	45.68

Assume the reference axis RF.

For the semicircle,

$$\text{Area} = \frac{0.785d^2}{2} = \frac{0.785 \times 4 \times 4}{2} = 6.28 \text{ sq in.}$$

x distance $= 0.576r = 0.576 \times 2 = 1.15$ in.*

$Ax = 6.28 \times 1.15 = 7.22$ cu in.

$I_{cg} = 0.110r^4 = 0.110 \times 2 \times 2 \times 2 \times 2 = 1.76$ in.^{4}

For the trapezoid,

$$\text{Area} = \frac{h(b + b_1)}{2} = \frac{4(4 + 2)}{2} = 12 \text{ sq in.}$$

Distance from c.g. axis to smaller base $= \dfrac{h(2b + b_1)}{3(b + b_1)}$

$$= \frac{4(2 \times 4 + 2)}{3(4 + 2)} = \frac{40}{18} = 2.22 \text{ in.}$$

x distance $= 2 + 4 - 2.22 = 3.78$ in.

$Ax = 12 \times 3.78 = 45.4$ cu in.

$$I_{cg} = \frac{h^3(b^2 + 4bb_1 + b_1{}^2)}{36(b + b_1)} = \frac{64(16 + 32 + 4)}{36 \times 6} = 15.4 \text{ in.}^4$$

For the compound section,

$$x_G = \frac{\Sigma \text{ area moments}}{\Sigma \text{ areas}} = \frac{52.62}{18.28} = 2.88 \text{ in.} \tag{65}$$

x_I (for semicircle) $= 2.88 - 1.15 = 1.73$ in.

x_I (for trapezoid) $= 3.78 - 2.88 = 0.90$ in.

Distance e from c.g. of compound section to line of action of forces $= 2.88 + 6 = 8.88$ in.

Calculation of safe load:

$$P = \frac{sAI}{I + Aec} = \frac{8000 \times 18.3 \times 45.7}{45.7 + 18.3 \times 8.88 \times 2.88}$$

$$= \frac{6,690,000}{45.7 + 468} = 13,000 \text{ lb} \tag{63}$$

$$P = \frac{sAI}{I - Aec'} = \frac{8000 \times 18.3 \times 45.7}{45.7 - 18.3 \times 8.88 \times 3.12}$$

$$= \frac{6,690,000}{45.7 - 507} = -14,500 \text{ lb} \tag{63a}$$

The hook can hold safely 13,000 lb.

* Formulas for this problem are in table on p. 215.

114. Compression and Bending—Columns. Short compression members (blocks or posts), as given in strength of materials, are assumed to be stressed equally throughout their cross sections. The stress is direct compression and its value is determined by

$$s = \frac{P}{A} \qquad (1)$$

Long compression members, on the other hand, will fail in bending as the applied axial load is increased. Roughly speaking, a compression member in which the length is more than ten times the smaller cross-sectional dimension should be considered as a column. This means that in the design of these members, due allowance must be made for bending stress as well as the direct compressive stress. Columns are examples of a very common form of combined stresses. The structural designer frequently must deal with column problems; the machine designer to a lesser extent. For example, it may be necessary to design the piston rods of a gas engine as columns.

The reason for the combined stresses induced in the crane hook, punch press, and the like are readily understood. That these members are subjected to direct tension is obvious, as well as the fact that bending is caused by the eccentricity of load application. But why, if a column is loaded axially, as seems to be the case in Fig. 140a, do the forces P give rise to bending, as shown exaggeratedly in Fig. 140b? Three factors are involved in the answer to this question.

1. Although a column may look straight, a slight bend in the length direction is always present before any load is applied.

2. It is impossible to make a column in which the material is perfectly homogeneous, that is, of the same quality and possessing the same properties throughout. Those parts which are less stiff will shorten more under load and cause bending.

3. The load will always bear somewhat more on one side than on the other, even though such is not the intention. The side that is more heavily loaded will shorten more and cause bending.

As the load on a column is increased, factors (2) and (3) cause the bending to increase. Then, as this process continues, a condition is brought about in which bending will continue until failure, without any further load application. The maximum load that a column can support without bending failure is called its *critical load*.

Similar to cases of tension and bending, discussed in the crane-hook type of problem, the stress of direct compression and the compression and tension of bending are added algebraically at any section to obtain the true stress at that section. The direct compressive stress is assumed to be equal throughout the column, but the bending stress would normally be greatest at the mid-section.[1] Hence, the dangerous section is AA of Fig. 140b. The compression caused by bending occurs on the inside of the bent column and the tension on the outside. The true stress on section AA, represented in Fig. 140c, varies in intensity, but, as is usual in columns, is compression at every point. The maximum stress on the dangerous section, here indicated by the arrow at the extreme right of Fig. 140c, for safety must not exceed

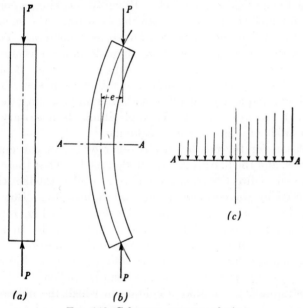

FIG. 140. Columns—stress analysis.

the design stress of the material. Again, in a fashion similar to the crane-hook analysis, let s_1 be the stress caused by direct compression, s_2 the maximum compressive bending stress, and s the total of the two. Then

$$s = s_1 + s_2$$

Also, as before,

$$s = \frac{P}{A} + \frac{Mc}{I}$$

and

$$\frac{P}{A} = s - \frac{Mc}{I}$$

where $M = Pe$, as in Fig. 140b. While the maximum stress on the

[1] Except in rigid frame structures.

dangerous section of a column must for safety be no greater than the design stress of the material, the *average* stress over the area must always be considerably less than the design-stress value. For any load the average stress is

$$s_{\mathrm{av}} = \frac{P}{A}$$

In the equation

$$\frac{P}{A} = s - \frac{Mc}{I}$$

we can see that in a column the average stress, P/A, is equal to the maximum stress minus the stress of bending. Column formulas follow this general pattern, but because of the fact that the extent of bending caused by the above-mentioned factors (1) to (3) is not known, the eccentricity e cannot be determined, and therefore further exact analysis is impossible. Column formulas are for these reasons based in part on theoretical analysis and in part on the results of experience (empirical).

The column formula developed by Johnson has found favor among machine designers within recent years. It is applicable to what is known as *intermediate columns*, that is, with values of the *slenderness ratio* (l/k) not in excess of 100.[1] This formula, where the entire right-hand side of the equation is a value of the average stress s_{av}, is

$$\frac{P}{A} = s\left[1 - \frac{s_e}{4C\pi^2 E}\left(\frac{l}{k}\right)^2\right] \tag{66}$$

where s = design stress of material (when safe load is to be calculated)
 s_e = elastic limit of material
 l = unsupported length of column, in.
 k = *least* radius of gyration and is equal to $\sqrt{I/A}$ (where I is the least moment of inertia about the neutral axis)
 C = a factor depending on method of securing the ends, as explained below
 E = modulus of elasticity of material

As in beams, columns with firmly fixed ends offer more resistance to bending than columns with ends secured otherwise, for example, those held with pins through the ends. A value of 4 for C is given for fixed-end and a value of 1 for pin-end columns. Intermediate conditions call for values in between.

Illustrative Example. The connecting rod of a steam locomotive is to be investigated as a column to determine the safe load. The cross

[1] For long columns, Euler's formula should be used.

section is circular with a diameter of 3 in. and the length of the rod is 60 in. Assume a design stress for the steel of 9000 psi in compression, elastic limit of 40,000 psi, and modulus of elasticity of 30,000,000. The connections are pinned ($C = 1$).

$$P = As\left[1 - \frac{s_e}{4C\pi^2E}\left(\frac{l}{k}\right)^2\right] \tag{66}$$

$$I = \frac{\pi d^4}{64}$$

$$k^2 = \frac{I}{A} = \frac{\pi d^4}{64} \times \frac{4}{\pi d^2} = \frac{d^2}{16} = \frac{9}{16} = 0.563 \text{ sq in.}$$

$$P = \frac{3.14 \times 3 \times 3 \times 9000}{4}$$

$$\times \left(1 - \frac{40,000 \times 60 \times 60}{4 \times 1 \times 3.14 \times 3.14 \times 30,000,000 \times 0.563}\right)$$

$$= 63,600(1 - 0.216) = 49,900 \text{ lb}$$

At times designers are confronted with problems of eccentrically loaded columns. This means that the column load is applied some

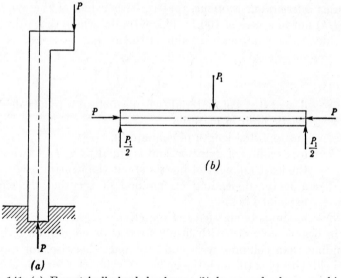

(b)

(a)

FIG. 141. (a) Eccentrically loaded column; (b) beam and column combined.

distance away from the center of gravity of the column cross section, as shown in Fig. 141a. The actual maximum stress in a column of this kind is the sum of the direct compression, the maximum compression induced by column bending, and compression induced by the new factor of bending because of the eccentricity of the load.　All these factors are considered in design.

Also at times a part is used both as a column and as a beam.　Figure 141b illustrates this case.　Again three factors must be considered, namely, direct compression, column bending, and beam bending.

115. Design of a Column of Circular Cross Section.　When dealing with columns of circular cross section, column formulas can be applied directly to determine the cross-sectional diameter required to support the given loads.　For this purpose it is found convenient to transform the equation somewhat.　As pointed out above, for a circular section,

$$A = \frac{\pi d^2}{4} \quad \text{and} \quad k^2 = \frac{d^2}{16}$$

Then in the Johnson formula,

$$P = As\left[1 - \frac{s_e}{4C\pi^2 E}\left(\frac{l}{k}\right)^2\right] \tag{66}$$

these substitutions are made.　By transformations and combinations, the equation eventually reduces to

$$d = \sqrt{\frac{4P}{\pi s} + \frac{4s_e l^2}{C\pi^2 E}} \tag{67}$$

Illustrative Example.　A steel rod, circular in cross section and 26 in. long, must support repeated axial reversing loads up to 10,000 lb as a part of a large printing press.　Calculate the required diameter, assuming s as 8000 psi, s_e as 45,000 psi, and E as 30,000,000.　The ends are partially fixed to make the value of C equal to 2.

$$d = \sqrt{\frac{4P}{\pi s} + \frac{4s_e l^2}{C\pi^2 E}} = \sqrt{\frac{4 \times 10,000}{3.14 \times 8000} + \frac{4 \times 45,000 \times 26 \times 26}{2 \times 3.14 \times 3.14 \times 30,000,000}}$$

$$= \sqrt{1.59 + 0.206} = \sqrt{1.80} = 1.34 \text{ in.} \tag{67}$$

Use $1\frac{3}{8}$ in. = 1.38 in.

Check to see if l/k is less than 100.

$$k = \frac{d}{4} = \frac{1.38}{4}$$

$$\frac{l}{k} = \frac{26 \times 4}{1.38} = 75.4$$

PROBLEMS

1. A crane hook is circular in cross section with a diameter of 2 in. at the dangerous section. The eccentricity (distance from the center of gravity of the dangerous section to the line of action of the forces) is 3 in. Calculate the maximum load that the hook can support safely. Use a design stress of 8000 psi for tension.

2. What is the maximum allowable eccentricity for a crane hook with a diameter of 1¼ in. at the dangerous section to support a load of 1000 lb. Assume a design stress of 9000 psi.

3. The compound section shown in the accompanying figure may be thought of as composed of an ellipse and a rectangle.

  a. Locate the center-of-gravity axes.

  b. Calculate the moments of inertia about these axes.

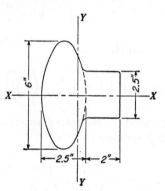

4. The cross section of the frame of a punch press where the bending is a maximum is the shape of a T. The top of the T is a rectangle 8 by 2 in. and the stem a rectangle 2½ by 4 in. with the 2½-in. side adjacent to the top. There is a distance of 8 in. from the nearest edge of the T to the punching force. Calculate the stress developed when the punch is punching 1-in.-diam holes from a brass plate ⅛ in. thick. The ultimate strength of the brass in shear is 30,000 psi.

5. A punch press has a dangerous section in the shape of a trapezoid. The base at which there is maximum tension is 6 in., the other base is 2 in., and the altitude is 5 in. There is a distance of 14 in. between the line of action of the punching force and the larger base. Calculate the safe load that can be applied to the punch, using a design stress of 5000 psi for the cast-iron frame.

6. The dangerous section of a crane hook may be considered to be made up of a 2- by 2-in. square and an isosceles triangle with a 2-in. base adjacent to the square and an altitude of 3 in. The distance from the line of action of the forces to the nearest side of the square is 5 in. Calculate the safe load when the limit stress is 32,800 lb and the factor of safety is 4.

7. A crane hook must be able to lift a load of 20,000 lb. The dangerous section is in the form of a semicircle with a radius of 1½ in. and an isosceles triangle with a base of 3 in. adjacent to the semicircle and an altitude of 5 in. The distance from the line of action of the forces to the nearest point of the dangerous section is 5 in. Make calculations to determine whether the stress developed is within the design-stress value of 10,000 psi.

8. The punch press of the accompanying figure has a tensile section modulus (I/c) at section xx of 53.5 cu in., a compressive section modulus of 28.9 cu in., and an area of 24.4 sq in. What is the diameter of the largest hole that should be punched from ¼-in. steel plate? Assume an ultimate strength of 45,000 psi for the steel in shear and a design stress of 6000 psi for the frame of the punch.

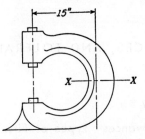

9. The cross section of a machine part is made up of a 2- by 4-in. rectangle and two 1-in.-radius semicircles on each of the smaller sides of the rectangle to make the shape of an oval. Calculate the least radius of gyration as a step in the investigation of the part as a column.

10. A certain cast-iron machine part, 1½ by 2½ in. in cross section and 2 ft 8 in. long is alternately stressed in tension and compression by axial loads. Calculate the safe load that the part can hold, assuming design stresses for tension and compression of 10,000 psi and an elastic limit of 60,000 psi. Consider the ends as partially fixed $(C = 2)$. The modulus of elasticity for cast iron is 15,000,000.

11. A compression member is elliptical in cross section with a 4-in. major axis, a 2½-in. minor axis, and a length of 3 ft 6 in. The design stress is 12,000 psi and the elastic limit is 36,000 psi. Calculate the safe load, assuming the ends to be fixed $(C = 4)$. The modulus of elasticity is 30,000,000.

12. The piston rod of a diesel engine has a diameter of 2 in. and is 28 in. long. The explosion transmits a maximum axial force of 22,500 lb to the rod. Calculate the maximum stress developed. Assume an elastic limit of 50,000 psi and that the ends are free. The modulus of elasticity is 30,000,000.

13. Calculate the required diameter for a steel compression member, 40 in. long, to hold an axial compressive load of 24,800 lb. The steel has an elastic limit of 50,000 psi and the design stress is 9000 psi. The ends are free and the modulus of elasticity is 30,000,000.

14. A pressure of 200 psi on a piston, 52 sq in. in area, is transmitted as an axial load to a piston rod, 2 ft 6 in. in length. Calculate the required diameter of the rod. Assume a design stress of 9000 psi and an elastic stress of 60,000 psi. The modulus of elasticity is 28,000,000. One end is pinned; one end is fixed $(C = 2)$.

CHAPTER 21

FITS, ALLOWANCES, AND TOLERANCES

116. Fits and Allowances—Types of Fits. Many of the various machine parts studied in the preceding chapters are intended to fit into or through other parts as assembled mechanisms. A shaft through the hub of a pulley, the teeth of mating gears, and screws in tapped holes are examples. In these chapters we have been interested mainly in the design of these parts for strength and economy and have given little thought to the problem of assembly. In this last chapter we shall discuss some phases of this problem.

The degree of tightness of contact between two bodies, one of which partially or completely envelops the other, is called *fit*. A loose fit requires a definite clearance between mating parts, as, for example, the clearance needed between the disk of a friction clutch and the splined shaft on which the disk slides. The clearance is known as the *allowance*. On the other hand, in some cases the fit must be so tight that the two parts will be able to transmit forces or torques through their contact surfaces. In this way keying is made unnecessary. An example may be found in the handwheels on the feed screws of some lathes. Fits, as given by the ASME, range from loose to very tight in eight stages.

Let us now consider the hub and shaft of Fig. 142. If the shaft diameter is slightly larger than the hole diameter, it is still possible, through the application of force, to fit the shaft into the hole. This is called a *force fit*. A similar result can be accomplished by heating the hub to expand the hole diameter beyond that of the shaft before assembling the parts. This is called a *shrink fit*. In both of these cases we say that there is a negative allowance or *interference* between the original diameters of hub and shaft.

Both force and shrink fits stretch the hole and compress the shaft, thereby inducing stresses of tension and compression, respectively. These stresses create normal forces of high magnitude on shaft and hole surfaces to prevent slippage when torques are transmitted between shaft and hub. For the same interference a shrink joint is able to transmit greater torque than a force joint, for the reason that abrasion takes place during the forcing operation and the interference is reduced.

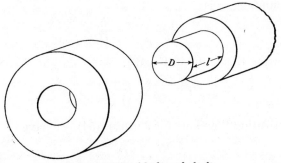

Fig. 142. Fit of hub and shaft.

117. Stresses in Shrink and Force Joints. Because of the fact that both of the mating parts change in size and are stressed in a force or shrink joint, the calculations for the value of the induced stress is quite complex. However, if it is assumed that the shaft diameter does not decrease (which condition is approached with large-diameter shafts and thin hubs), then the hub only is stressed, and calculations become relatively simple.

Such cases are analogous to thin-walled cylinders, described in Chap. 17, and may be treated in the same way. Instead of a fluid pressure on the inside of a tank or boiler, the shaft is creating a pressure this time on the inside of the hub. As developed in Sec. 89, the tensile stress in the boiler wall is

$$s = \frac{pD}{2t} \tag{54}$$

When this equation is applied to a shrink or force joint,

$s =$ tensile stress in hub
$D =$ shaft diameter
$t =$ hub thickness
$p =$ normal pressure of hub on shaft

With increased interference, the hub tensile stress also increases, since more interference means more stretching of the hub and

$$s = E\delta \tag{3}$$

for values at or below the elastic limit. If i represents the interference (difference between shaft diameter and hub diameter before fitting), then

$$\text{Total circumferential stretch of hole} = \pi i$$

But the circumference is πD, and therefore the stretch per linear inch of hole circumference is

$$\delta = \frac{\pi i}{\pi D} = \frac{i}{D}$$

By substituting this value in Eq. (3),

$$s = \frac{Ei}{D} \tag{68}$$

Again substituting the value for s in Eq. (54), we have

$$\frac{Ei}{D} = \frac{pD}{2t}$$

and

$$p = \frac{2Eit}{D^2} \tag{69}$$

With this equation the normal pressure of hub on shaft can be calculated when the interference is known, provided that the hub is not stressed above the elastic limit.

118. Resisting Torques of Shrink and Force Joints. What resisting torque value can be developed by a shrink or force joint? This value depends on the frictional resistance between hub and shaft and on the shaft radius. In symbols,

$$T = Fr = \frac{FD}{2}$$

where F is the frictional force. Also as previously developed,

$$F = Nf \tag{38}$$

The normal force N is the product of the normal pressure p and the area of contact.

$$N = p \times A = p \times l\pi D$$

As shown in Sec. 117,

$$p = \frac{2Eit}{D^2} \tag{69}$$

Making these various substitutions in the torque equation, we have

$$T = p \times A \times f \times r = \frac{2Eit}{D^2} \times l\pi D \times f \times \frac{D}{2}$$

and, by cancellation,

$$T = Eitl\pi f \tag{70}$$

Although it was pointed out earlier that shrink joints are actually stronger than force joints, the equations developed here are used for both.

For force joints the force necessary to assemble the parts is again that needed to overcome friction. This force is small at first but increases in proportion to the area of contact and reaches a maximum just before assembly is completed. Again using the example of a hub and shaft, we note that the direction of the frictional force resisting the pressing of the parts together is axial, whereas the direction of the frictional force of the resisting torque is circumferential. However, the *value* of the two is the same, and since

$$T = \frac{FD}{2}$$

by substitution in Eq. (70) and transformation, we have

$$F = \frac{2Eitl\pi f}{D} \tag{71}$$

Illustrative Example. A steel collar, $\frac{3}{8}$ in. thick and $1\frac{1}{2}$ in. long, is to be forced around a 2-in.-diam shaft. The stress in the collar must not exceed 20,000 psi. Calculate:

a. The interference between shaft and collar

b. The torque that the joint can develop

The modulus of elasticity is 30,000,000 and the coefficient of friction between the parts is 0.25.

a.

$$s = \frac{Ei}{D} \tag{68}$$

$$i = \frac{sD}{E} = \frac{20,000 \times 2}{30,000,000} = 0.00133, \text{ or } 0.0013 \text{ in.}$$

b.

$$T = Eitl\pi f$$
$$= 30,000,000 \times 0.0013 \times 0.375 \times 1.5 \times 3.14 \times 0.25$$
$$= 17,200 \text{ lb-in.} \tag{70}$$

119. Tolerances. When we speak of a 1-in.-diam shaft, a $\frac{3}{4}$-in. face of gear tooth, or a $\frac{1}{2}$-in. square key, we mean that these are the dimensions of the parts within narrow limits. The amount by which each part may depart from its nominal dimension of 1 in., $\frac{3}{4}$ in., or

½ in. depends on the job requirements. Since a shaft must fit into bearings and support gears or pulleys, not much variation between actual size and nominal size is permissible. The same applies to the key which must be fitted into the keyway. On the other hand, it would probably make little difference if the tooth face were $2\frac{3}{32}$ in. or $2\frac{5}{32}$ in. instead of $\frac{3}{4}$ in.

Tolerance can be defined as the permissible margin of error in the making of a part. The smaller the tolerance, the more precise must be the work of the man in the shop, and consequently the more expensive is the production of the part. Tolerances for all parts as well as dimensions should be noted on drawings before they leave the design office.

For nonmating parts, such as the head of a bolt or the thickness of the leaf of a spring, the dimensions are permitted to run slightly over or under the nominal dimensions, for example, ±0.005 in. However, when the part has been designed for strength, caution must be taken that the minus tolerance does not appreciably weaken the part.

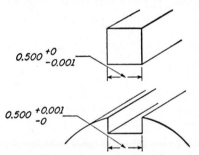

FIG. 143. Tolerances for key and keyway.

For mating parts the problem of setting the proper tolerance is more complex. Let us take an example of a square or rectangular key to be fitted into a key seat in a shaft (Fig. 143). If the nominal size of key and key seat is ½ in., the key cannot be more than ½ in. or the key seat less than ½ in. or there will be interference. Therefore, the tolerance specified for the key is minus and that for the key seat is plus. Also, in order to have a snug fit, these tolerances should not be large. A reasonable tolerance would be −0.001 in. for the width of key and +0.001 in. for the width of key seat.

PROBLEMS

1. A steel tire is to be bound on a locomotive wheel 6 ft in diameter by means of a shrink fit. There is an interference of 0.025 in. between wheel and tire diameters.   Calculate the stress that will be induced in the tire, assuming the modulus of elasticity of steel to be 29,000,000.

2. A split hub is to be held tight to a shaft by two steel rings shrunk into shoulders on each side of the hub.   The cross section of each ring is ¼ in. square and the inside diameter is 3.035 in.   The stress in the rings should not exceed 24,000 psi.   Calculate the maximum diameter of shoulder, assuming a modulus of elasticity of 29,000,000.

3. A bronze hub, ½ in. thick and 2 in. long, is to be pressed over a 2½-in.-diam cast-iron shaft.   The exact diameter of the shaft is 2.506 in., and the exact inside diameter of the hub is 2.498 in.   Calculate:

    a. The force needed to make the fit

    b. The stress induced in the hub

    c. The resisting torque

Assume a modulus of elasticity of 4,500,000 for the bronze and a coefficient of friction between the two parts of 0.25.

4. The accompanying figure shows a shrink joint at the rim of a flywheel. Each link is 1¼ in. square in cross section, and there are two links at each joint. When the wheel is rotating at maximum speed, the centrifugal force causes a circumferential tension of 12,000 lb in the rim.   The stress in the links is not to exceed 30,000 psi.   Calculate the maximum interference, assuming a modulus of elasticity of 28,000,000.

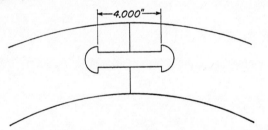

5. The link steel of Prob. 4 lengthens 0.0000067 in. for each inch of length per degree Fahrenheit.   How many degrees must the temperature be raised to make the fit?

6. The hub of a crank for a milling-machine vise is to be pressed on to the ¾-in.-diam shank at the end of the vise screw.   When assembled, the crank handle would be 3 in. distant perpendicularly from the screw axis.   The joint must be able to withstand a force of 500 lb applied to the handle.   Calculate the length and thickness of the hub, making the length equal to three times the thickness.   Assume a coefficient of friction of 0.20, a modulus of elasticity of 30,000,000, and a design stress in tension of 20,000 psi.

7. A collar thrust bearing is to be made by pressing three collars with outside diameters of approximately 5¼ in. and lengths of 1 in. around a shaft with a diameter of 3.725 in.   The allowable bearing pressure is 90 psi.   It is specified that the force needed to prevent slippage between each collar and the shaft must be at least ten times the thrust on the collar.   Calculate the maximum inside diameters of the collars.   Use a modulus of elasticity of 30,000,000 and a coefficient of friction of 0.20.

APPENDIX

AREAS AND MOMENTS OF INERTIA ABOUT CENTER-OF-GRAVITY AXES OF SIMPLE GEOMETRIC SECTIONS

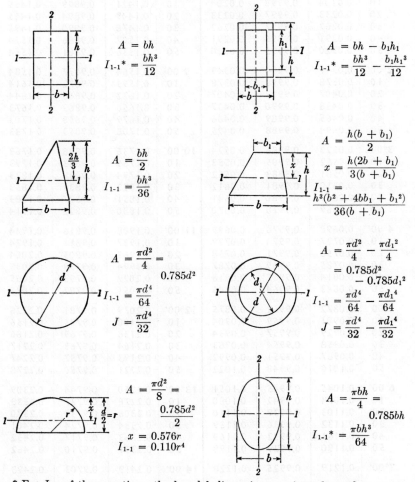

$$A = bh$$
$$I_{1\text{-}1}{}^* = \frac{bh^3}{12}$$

$$A = bh - b_1h_1$$
$$I_{1\text{-}1}{}^* = \frac{bh^3}{12} - \frac{b_1h_1{}^3}{12}$$

$$A = \frac{bh}{2}$$
$$I_{1\text{-}1} = \frac{bh^3}{36}$$

$$A = \frac{h(b + b_1)}{2}$$
$$x = \frac{h(2b + b_1)}{3(b + b_1)}$$
$$I_{1\text{-}1} = \frac{h^3(b^2 + 4bb_1 + b_1{}^2)}{36(b + b_1)}$$

$$A = \frac{\pi d^2}{4} = 0.785d^2$$
$$I_{1\text{-}1} = \frac{\pi d^4}{64}$$
$$J = \frac{\pi d^4}{32}$$

$$A = \frac{\pi d^2}{4} - \frac{\pi d_1{}^2}{4}$$
$$= 0.785d^2 - 0.785d_1{}^2$$
$$I_{1\text{-}1} = \frac{\pi d^4}{64} - \frac{\pi d_1{}^4}{64}$$
$$J = \frac{\pi d^4}{32} - \frac{\pi d_1{}^4}{32}$$

$$A = \frac{\pi d^2}{8} = \frac{0.785d^2}{2}$$
$$x = 0.576r$$
$$I_{1\text{-}1} = 0.110r^4$$

$$A = \frac{\pi bh}{4} = 0.785bh$$
$$I_{1\text{-}1}{}^* = \frac{\pi bh^3}{64}$$

* For $I_{2\text{-}2}$ of these sections, the b and h dimensions are interchanged.

215

ELEMENTS OF MACHINE DESIGN

TABLE OF TRIGONOMETRIC RATIOS

Angle	Sine	Cosine	Tangent	Angle	Sine	Cosine	Tangent
0°00′	0.0000	1.0000	0.0000	7°00′	0.1219	0.9925	0.1228
10	0.0029		0.0029	10	0.1248	0.9922	0.1257
20	0.0058	approx.	0.0058	20	0.1276	0.9918	0.1287
30	0.0087	1.	0.0087	30	0.1305	0.9914	0.1317
40	0.0116	0.9999	0.0116	40	0.1334	0.9911	0.1346
50	0.0145	0.9999	0.0145	50	0.1363	0.9907	0.1376
1°00′	0.0175	0.9998	0.0175	8°00′	0.1392	0.9903	0.1405
10	0.0204	0.9998	0.0204	10	0.1421	0.9899	0.1435
20	0.0233	0.9997	0.0233	20	0.1449	0.9894	0.1465
30	0.0262	0.9997	0.0262	30	0.1478	0.9890	0.1495
40	0.0291	0.9996	0.0291	40	0.1507	0.9886	0.1524
50	0.0320	0.9995	0.0320	50	0.1536	0.9881	0.1554
2°00′	0.0349	0.9994	0.0349	9°00′	0.1564	0.9877	0.1584
10	0.0378	0.9993	0.0378	10	0.1593	0.9872	0.1614
20	0.0407	0.9992	0.0407	20	0.1622	0.9868	0.1644
30	0.0436	0.9990	0.0437	30	0.1650	0.9863	0.1673
40	0.0465	0.9989	0.0466	40	0.1679	0.9858	0.1703
50	0.0494	0.9988	0.0495	50	0.1708	0.9853	0.1733
3°00′	0.0523	0.9986	0.0524	10°00′	0.1736	0.9848	0.1763
10	0.0552	0.9985	0.0553	10	0.1765	0.9843	0.1793
20	0.0581	0.9983	0.0582	20	0.1794	0.9838	0.1823
30	0.0610	0.9981	0.0612	30	0.1822	0.9833	0.1853
40	0.0640	0.9980	0.0641	40	0.1851	0.9827	0.1883
50	0.0669	0.9978	0.0670	50	0.1880	0.9822	0.1914
4°00′	0.0698	0.9976	0.0699	11°00′	0.1908	0.9816	0.1944
10	0.0727	0.9974	0.0729	10	0.1937	0.9811	0.1974
20	0.0756	0.9971	0.0758	20	0.1965	0.9805	0.2004
30	0.0785	0.9969	0.0787	30	0.1994	0.9799	0.2035
40	0.0814	0.9967	0.0816	40	0.2022	0.9793	0.2065
50	0.0843	0.9964	0.0846	50	0.2051	0.9787	0.2095
5°00′	0.0872	0.9962	0.0875	12°00′	0.2079	0.9781	0.2126
10	0.0901	0.9959	0.0904	10	0.2108	0.9775	0.2156
20	0.0929	0.9957	0.0934	20	0.2136	0.9769	0.2186
30	0.0958	0.9954	0.0963	30	0.2164	0.9763	0.2217
40	0.0987	0.9951	0.0992	40	0.2193	0.9757	0.2247
50	0.1016	0.9948	0.1022	50	0.2221	0.9750	0.2278
6°00′	0.1045	0.9945	0.1051	13°00′	0.2250	0.9744	0.2309
10	0.1074	0.9942	0.1080	10	0.2278	0.9737	0.2339
20	0.1103	0.9939	0.1110	20	0.2306	0.9730	0.2370
30	0.1132	0.9936	0.1139	30	0.2334	0.9724	0.2401
40	0.1161	0.9932	0.1169	40	0.2363	0.9717	0.2432
50	0.1190	0.9929	0.1198	50	0.2391	0.9710	0.2462
7°00′	0.1219	0.9925	0.1228	14°00′	0.2419	0.9703	0.2493

TABLE OF TRIGONOMETRIC RATIOS (*Cont.*)

Angle	Sine	Cosine	Tangent	Angle	Sine	Cosine	Tangent
14°00′	0.2419	0.9703	0.2493	21°00′	0.3584	0.9336	0.3839
10	0.2447	0.9696	0.2524	10	0.3611	0.9325	0.3872
20	0.2476	0.9689	0.2555	20	0.3638	0.9315	0.3906
30	0.2504	0.9681	0.2586	30	0.3665	0.9304	0.3939
40	0.2532	0.9674	0.2617	40	0.3692	0.9293	0.3973
50	0.2560	0.9667	0.2648	50	0.3719	0.9283	0.4006
15°00′	0.2588	0.9659	0.2679	22°00′	0.3746	0.9272	0.4040
10	0.2616	0.9652	0.2711	10	0.3773	0.9261	0.4074
20	0.2644	0.9644	0.2742	20	0.3800	0.9250	0.4108
30	0.2672	0.9636	0.2773	30	0.3827	0.9239	0.4142
40	0.2700	0.9628	0.2805	40	0.3854	0.9228	0.4176
50	0.2728	0.9621	0.2836	50	0.3881	0.9216	0.4210
16°00′	0.2756	0.9613	0.2867	23°00′	0.3907	0.9205	0.4245
10	0.2784	0.9605	0.2899	10	0.3934	0.9194	0.4279
20	0.2812	0.9596	0.2931	20	0.3961	0.9182	0.4314
30	0.2840	0.9588	0.2962	30	0.3987	0.9171	0.4348
40	0.2868	0.9580	0.2994	40	0.4014	0.9159	0.4383
50	0.2896	0.9572	0.3026	50	0.4041	0.9147	0.4417
17°00′	0.2924	0.9563	0.3057	24°00′	0.4067	0.9135	0.4452
10	0.2952	0.9555	0.3089	10	0.4094	0.9124	0.4487
20	0.2979	0.9546	0.3121	20	0.4120	0.9112	0.4522
30	0.3007	0.9537	0.3153	30	0.4147	0.9100	0.4557
40	0.3035	0.9528	0.3185	40	0.4173	0.9088	0.4592
50	0.3062	0.9520	0.3217	50	0.4200	0.9075	0.4628
18°00′	0.3090	0.9511	0.3249	25°00′	0.4226	0.9063	0.4663
10	0.3118	0.9502	0.3281	10	0.4253	0.9051	0.4699
20	0.3145	0.9492	0.3314	20	0.4279	0.9038	0.4734
30	0.3173	0.9483	0.3346	30	0.4305	0.9026	0.4770
40	0.3201	0.9474	0.3378	40	0.4331	0.9013	0.4806
50	0.3228	0.9465	0.3411	50	0.4358	0.9001	0.4841
19°00′	0.3256	0.9455	0.3443	26°00′	0.4384	0.8988	0.4877
10	0.3283	0.9446	0.3476	10	0.4410	0.8975	0.4913
20	0.3311	0.9436	0.3508	20	0.4436	0.8962	0.4950
30	0.3338	0.9426	0.3541	30	0.4462	0.8949	0.4986
40	0.3365	0.9417	0.3574	40	0.4488	0.8936	0.5022
50	0.3393	0.9407	0.3607	50	0.4514	0.8923	0.5059
20°00′	0.3420	0.9397	0.3640	27°00′	0.4540	0.8910	0.5095
10	0.3448	0.9387	0.3673	10	0.4566	0.8897	0.5132
20	0.3475	0.9377	0.3706	20	0.4592	0.8884	0.5169
30	0.3502	0.9367	0.3739	30	0.4617	0.8870	0.5206
40	0.3529	0.9356	0.3772	40	0.4643	0.8857	0.5243
50	0.3557	0.9346	0.3805	50	0.4669	0.8843	0.5280
21°00′	0.3584	0.9336	0.3839	28°00′	0.4695	0.8829	0.5317

TABLE OF TRIGONOMETRIC RATIOS (*Cont.*)

Angle	Sine	Cosine	Tangent	Angle	Sine	Cosine	Tangent
28°00′	0.4695	0.8829	0.5317	35°00′	0.5736	0.8192	0.7002
10	0.4720	0.8816	0.5354	10	0.5760	0.8175	0.7046
20	0.4746	0.8802	0.5392	20	0.5783	0.8158	0.7089
30	0.4772	0.8788	0.5430	30	0.5807	0.8141	0.7133
40	0.4797	0.8774	0.5467	40	0.5831	0.8124	0.7177
50	0.4823	0.8760	0.5505	50	0.5854	0.8107	0.7221
29°00′	0.4848	0.8746	0.5543	36°00′	0.5878	0.8090	0.7265
10	0.4874	0.8732	0.5581	10	0.5901	0.8073	0.7310
20	0.4899	0.8718	0.5619	20	0.5925	0.8056	0.7355
30	0.4924	0.8704	0.5658	30	0.5948	0.8039	0.7400
40	0.4950	0.8689	0.5696	40	0.5972	0.8021	0.7445
50	0.4975	0.8675	0.5735	50	0.5995	0.8004	0.7490
30°00′	0.5000	0.8660	0.5774	37°00′	0.6018	0.7986	0.7536
10	0.5025	0.8646	0.5812	10	0.6041	0.7969	0.7581
20	0.5050	0.8631	0.5851	20	0.6065	0.7951	0.7627
30	0.5075	0.8616	0.5890	30	0.6088	0.7934	0.7673
40	0.5100	0.8601	0.5930	40	0.6111	0.7916	0.7720
50	0.5125	0.8587	0.5969	50	0.6134	0.7898	0.7766
31°00′	0.5150	0.8572	0.6009	38°00′	0.6157	0.7880	0.7813
10	0.5175	0.8557	0.6048	10	0.6180	0.7862	0.7860
20	0.5200	0.8542	0.6088	20	0.6202	0.7844	0.7907
30	0.5225	0.8526	0.6128	30	0.6225	0.7826	0.7954
40	0.5250	0.8511	0.6168	40	0.6248	0.7808	0.8002
50	0.5275	0.8496	0.6208	50	0.6271	0.7790	0.8050
32°00′	0.5299	0.8480	0.6249	39°00′	0.6293	0.7771	0.8098
10	0.5324	0.8465	0.6289	10	0.6316	0.7753	0.8146
20	0.5348	0.8450	0.6330	20	0.6338	0.7735	0.8195
30	0.5373	0.8434	0.6371	30	0.6361	0.7716	0.8243
40	0.5398	0.8418	0.6412	40	0.6383	0.7698	0.8292
50	0.5422	0.8403	0.6453	50	0.6406	0.7679	0.8342
33°00′	0.5446	0.8387	0.6494	40°00′	0.6428	0.7660	0.8391
10	0.5471	0.8371	0.6536	10	0.6450	0.7642	0.8441
20	0.5495	0.8355	0.6577	20	0.6472	0.7623	0.8491
30	0.5519	0.8339	0.6619	30	0.6494	0.7604	0.8541
40	0.5544	0.8323	0.6661	40	0.6517	0.7585	0.8591
50	0.5568	0.8307	0.6703	50	0.6539	0.7566	0.8642
34°00′	0.5592	0.8290	0.6745	41°00′	0.6561	0.7547	0.8693
10	0.5616	0.8274	0.6788	10	0.6583	0.7528	0.8744
20	0.5640	0.8258	0.6830	20	0.6604	0.7509	0.8796
30	0.5664	0.8241	0.6873	30	0.6626	0.7490	0.8847
40	0.5688	0.8225	0.6916	40	0.6648	0.7470	0.8899
50	0.5712	0.8208	0.6959	50	0.6670	0.7451	0.8952
35°00′	0.5736	0.8192	0.7002	42°00′	0.6691	0.7431	0.9004

TABLE OF TRIGONOMETRIC RATIOS (*Cont.*)

Angle	Sine	Cosine	Tangent	Angle	Sine	Cosine	Tangent
42°00′	0.6691	0.7431	0.9004	49°00′	0.7547	0.6561	1.1504
10	0.6713	0.7412	0.9057	10	0.7566	0.6539	1.1571
20	0.6734	0.7392	0.9110	20	0.7585	0.6517	1.1640
30	0.6756	0.7373	0.9163	30	0.7604	0.6494	1.1708
40	0.6777	0.7353	0.9217	40	0.7623	0.6472	1.1778
50	0.6799	0.7333	0.9271	50	0.7642	0.6450	1.1847
43°00′	0.6820	0.7314	0.9325	50°00′	0.7660	0.6428	1.1918
10	0.6841	0.7294	0.9380	10	0.7679	0.6406	1.1988
20	0.6862	0.7274	0.9435	20	0.7698	0.6383	1.2059
30	0.6884	0.7254	0.9490	30	0.7716	0.6361	1.2131
40	0.6905	0.7234	0.9545	40	0.7735	0.6338	1.2203
50	0.6926	0.7214	0.9601	50	0.7753	0.6316	1.2276
44°00′	0.6947	0.7193	0.9657	51°00′	0.7771	0.6293	1.2349
10	0.6967	0.7173	0.9713	10	0.7790	0.6271	1.2423
20	0.6988	0.7153	0.9770	20	0.7808	0.6248	1.2497
30	0.7009	0.7133	0.9827	30	0.7826	0.6225	1.2572
40	0.7030	0.7112	0.9884	40	0.7844	0.6202	1.2647
50	0.7050	0.7092	0.9942	50	0.7862	0.6180	1.2723
45°00′	0.7071	0.7071	1.0000	52°00′	0.7880	0.6157	1.2799
10	0.7092	0.7050	1.0058	10	0.7898	0.6134	1.2876
20	0.7112	0.7030	1.0117	20	0.7916	0.6111	1.2954
30	0.7133	0.7009	1.0176	30	0.7934	0.6088	1.3032
40	0.7153	0.6988	1.0235	40	0.7951	0.6065	1.3111
50	0.7173	0.6967	1.0295	50	0.7969	0.6041	1.3190
46°00′	0.7193	0.6947	1.0355	53°00′	0.7986	0.6018	1.3270
10	0.7214	0.6926	1.0416	10	0.8004	0.5995	1.3351
20	0.7234	0.6905	1.0477	20	0.8021	0.5972	1.3432
30	0.7254	0.6884	1.0538	30	0.8039	0.5948	1.3514
40	0.7274	0.6862	1.0599	40	0.8056	0.5925	1.3597
50	0.7294	0.6841	1.0661	50	0.8073	0.5901	1.3680
47°00′	0.7314	0.6820	1.0724	54°00′	0.8090	0.5878	1.3764
10	0.7333	0.6799	1.0786	10	0.8107	0.5854	1.3848
20	0.7353	0.6777	1.0850	20	0.8124	0.5831	1.3934
30	0.7373	0.6756	1.0913	30	0.8141	0.5807	1.4019
40	0.7392	0.6734	1.0977	40	0.8158	0.5783	1.4106
50	0.7412	0.6713	1.1041	50	0.8175	0.5760	1.4193
48°00′	0.7431	0.6691	1.1106	55°00′	0.8192	0.5736	1.4281
10	0.7451	0.6670	1.1171	10	0.8208	0.5712	1.4370
20	0.7470	0.6648	1.1237	20	0.8225	0.5688	1.4460
30	0.7490	0.6626	1.1303	30	0.8241	0.5664	1.4550
40	0.7509	0.6604	1.1369	40	0.8258	0.5640	1.4641
50	0.7528	0.6583	1.1436	50	0.8274	0.5616	1.4733
49°00′	0.7547	0.6561	1.1504	56°00′	0.8290	0.5592	1.4826

TABLE OF TRIGONOMETRIC RATIOS (*Cont.*)

Angle	Sine	Cosine	Tangent	Angle	Sine	Cosine	Tangent
56°00′	0.8290	0.5592	1.4826	63°00′	0.8910	0.4540	1.9626
10	0.8307	0.5568	1.4919	10	0.8923	0.4514	1.9768
20	0.8323	0.5544	1.5013	20	0.8936	0.4488	1.9912
30	0.8339	0.5519	1.5108	30	0.8949	0.4462	2.0057
40	0.8355	0.5495	1.5204	40	0.8962	0.4436	2.0204
50	0.8371	0.5471	1.5301	50	0.8975	0.4410	2.0353
57°00′	0.8387	0.5446	1.5399	64°00′	0.8988	0.4384	2.0503
10	0.8403	0.5422	1.5497	10	0.9001	0.4358	2.0655
20	0.8418	0.5398	1.5597	20	0.9013	0.4331	2.0809
30	0.8434	0.5373	1.5697	30	0.9026	0.4305	2.0965
40	0.8450	0.5348	1.5798	40	0.9038	0.4279	2.1123
50	0.8465	0.5324	1.5900	50	0.9051	0.4253	2.1283
58°00′	0.8480	0.5299	1.6003	65°00′	0.9063	0.4226	2.1445
10	0.8496	0.5275	1.6107	10	0.9075	0.4200	2.1609
20	0.8511	0.5250	1.6212	20	0.9088	0.4173	2.1175
30	0.8526	0.5225	1.6319	30	0.9100	0.4147	2.1943
40	0.8542	0.5200	1.6426	40	0.9112	0.4120	2.2113
50	0.8557	0.5175	1.6534	50	0.9124	0.4094	2.2286
59°00′	0.8572	0.5150	1.6643	66°00′	0.9135	0.4067	2.2460
10	0.8587	0.5125	1.6753	10	0.9147	0.4041	2.2637
20	0.8601	0.5100	1.6864	20	0.9159	0.4014	2.2817
30	0.8616	0.5075	1.6977	30	0.9171	0.3987	2.2998
40	0.8631	0.5050	1.7090	40	0.9182	0.3961	2.3183
50	0.8646	0.5025	1.7205	50	0.9194	0.3934	2.3369
60°00′	0.8660	0.5000	1.7321	67°00′	0.9205	0.3907	2.3559
10	0.8675	0.4975	1.7437	10	0.9216	0.3881	2.3750
20	0.8689	0.4950	1.7556	20	0.9228	0.3854	2.3945
30	0.8704	0.4924	1.7675	30	0.9239	0.3827	2.4142
40	0.8718	0.4899	1.7796	40	0.9250	0.3800	2.4342
50	0.8732	0.4874	1.7917	50	0.9261	0.3773	2.4545
61°00′	0.8746	0.4848	1.8040	68°00′	0.9272	0.3746	2.4751
10	0.8760	0.4823	1.8165	10	0.9283	0.3719	2.4960
20	0.8774	0.4797	1.8291	20	0.9293	0.3692	2.5172
30	0.8788	0.4772	1.8418	30	0.9304	0.3665	2.5386
40	0.8802	0.4746	1.8546	40	0.9315	0.3638	2.5605
50	0.8816	0.4720	1.8676	50	0.9325	0.3611	2.5826
62°00′	0.8829	0.4695	1.8807	69°00′	0.9336	0.3584	2.6051
10	0.8843	0.4669	1.8940	10	0.9346	0.3557	2.6279
20	0.8857	0.4643	1.9074	20	0.9356	0.3529	2.6511
30	0.8870	0.4617	1.9210	30	0.9367	0.3502	2.6746
40	0.8884	0.4592	1.9347	40	0.9377	0.3475	2.6985
50	0.8897	0.4566	1.9486	50	0.9387	0.3448	2.7228
63°00′	0.8910	0.4540	1.9626	70°00′	0.9397	0.3420	2.7475

TABLE OF TRIGONOMETRIC RATIOS *(Cont.)*

Angle	Sine	Cosine	Tangent	Angle	Sine	Cosine	Tangent
70°00′	0.9397	0.3420	2.7475	77°00′	0.9744	0.2250	4.3315
10	0.9407	0.3393	2.7725	10	0.9750	0.2221	4.3897
20	0.9417	0.3365	2.7980	20	0.9757	0.2193	4.4494
30	0.9426	0.3338	2.8239	30	0.9763	0.2164	4.5107
40	0.9436	0.3311	2.8502	40	0.9769	0.2136	4.5736
50	0.9446	0.3283	2.8770	50	0.9775	0.2108	4.6382
71°00′	0.9455	0.3256	2.9042	78°00′	0.9781	0.2079	4.7046
10	0.9465	0.3228	2.9319	10	0.9787	0.2051	4.7729
20	0.9474	0.3201	2.9600	20	0.9793	0.2022	4.8430
30	0.9483	0.3173	2.9887	30	0.9799	0.1994	4.9152
40	0.9492	0.3145	3.0178	40	0.9805	0.1965	4.9894
50	0.9502	0.3118	3.0475	50	0.9811	0.1937	5.0658
72°00′	0.9511	0.3090	3.0777	79°00′	0.9816	0.1908	5.1446
10	0.9520	0.3062	3.1084	10	0.9822	0.1880	5.2257
20	0.9528	0.3035	3.1397	20	0.9827	0.1851	5.3093
30	0.9537	0.3007	3.1716	30	0.9833	0.1822	5.3955
40	0.9546	0.2979	3.2041	40	0.9838	0.1794	5.4845
50	0.9555	0.2952	3.2371	50	0.9843	0.1765	5.5764
73°00′	0.9563	0.2924	3.2709	80°00′	0.9848	0.1736	5.6713
10	0.9572	0.2896	3.3052	10	0.9853	0.1708	5.7694
20	0.9580	0.2868	3.3402	20	0.9858	0.1679	5.8708
30	0.9588	0.2840	3.3759	30	0.9863	0.1650	5.9758
40	0.9596	0.2812	3.4124	40	0.9868	0.1622	6.0844
50	0.9605	0.2784	3.4495	50	0.9872	0.1593	6.1970
74°00′	0.9613	0.2756	3.4874	81°00′	0.9877	0.1564	6.3138
10	0.9621	0.2728	3.5261	10	0.9881	0.1536	6.4348
20	0.9628	0.2700	3.5656	20	0.9886	0.1507	6.5606
30	0.9636	0.2672	3.6059	30	0.9890	0.1478	6.6912
40	0.9644	0.2644	3.6470	40	0.9894	0.1449	6.8269
50	0.9652	0.2616	3.6891	50	0.9899	0.1421	6.9682
75°00′	0.9659	0.2588	3.7321	82°00′	0.9903	0.1392	7.1154
10	0.9667	0.2560	3.7760	10	0.9907	0.1363	7.2687
20	0.9674	0.2532	3.8208	20	0.9911	0.1334	7.4287
30	0.9681	0.2504	3.8667	30	0.9914	0.1305	7.5958
40	0.9689	0.2476	3.9136	40	0.9918	0.1276	7.7704
50	0.9696	0.2447	3.9617	50	0.9922	0.1248	7.9530
76°00′	0.9703	0.2419	4.0108	83°00′	0.9925	0.1219	8.1443
10	0.9710	0.2391	4.0611	10	0.9929	0.1190	8.3450
20	0.9717	0.2363	4.1126	20	0.9932	0.1161	8.5555
30	0.9724	0.2334	4.1653	30	0.9936	0.1132	8.7769
40	0.9730	0.2306	4.2193	40	0.9939	0.1103	9.0098
50	0.9737	0.2278	4.2747	50	0.9942	0.1074	9.2553
77°00′	0.9744	0.2250	4.3315	84°00′	0.9945	0.1045	9.5144

TABLE OF TRIGONOMETRIC RATIOS (*Cont.*)

Angle	Sine	Cosine	Tangent	Angle	Sine	Cosine	Tangent
84°00′	0.9945	0.1045	9.5144	87°00′	0.9986	0.0523	19.081
10	0.9948	0.1016	9.7882	10	0.9988	0.0494	20.206
20	0.9951	0.0987	10.078	20	0.9989	0.0465	21.470
30	0.9954	0.0958	10.385	30	0.9990	0.0436	22.904
40	0.9957	0.0929	10.712	40	0.9992	0.0407	24.542
50	0.9959	0.0901	11.059	50	0.9993	0.0378	26.432
85°00′	0.9962	0.0872	11.430	88°00′	0.9994	0.0349	28.636
10	0.9964	0.0843	11.826	10	0.9995	0.0320	31.242
20	0.9967	0.0814	12.251	20	0.9996	0.0291	34.368
30	0.9969	0.0785	12.706	30	0.9997	0.0262	38.188
40	0.9971	0.0756	13.197	40	0.9997	0.0233	42.964
50	0.9974	0.0727	13.727	50	0.9998	0.0204	49.104
86°00′	0.9976	0.0698	14.301	89°00′	0.9998	0.0175	57.290
10	0.9978	0.0669	14.924	10	0.9999	0.0145	68.750
20	0.9980	0.0640	15.605	20	0.9999	0.0116	85.940
30	0.9981	0.0610	16.350	30	approx. 1.	0.0087	114.59
40	0.9983	0.0581	17.169	40		0.0058	171.89
50	0.9985	0.0552	18.075	50		0.0029	343.77
87°00′	0.9986	0.0523	19.081	90°00′	1.0000	0.0000	Infinite

BIBLIOGRAPHY

PROPERTIES OF MATERIALS AND INDUSTRIAL PROCESSES

American Society for Metals: "Metals Handbook," Cleveland, 1948. (The most intensive, comprehensive, and authoritative book in this field.)

Camp, J. M., and C. B. Francis: "The Making, Shaping, and Treating of Steel," United States Steel Corporation, Pittsburgh, 1951. (An encyclopedic book. Excellent for industrial processes of steel.)

Coonan, F. L.: "Principles of Physical Metallurgy," Harper & Brothers, New York, 1943. (For a more detailed and advanced study of theory.)

Gillett, H. W.: "The Behavior of Engineering Metals," John Wiley & Sons, Inc., New York, 1951. (A comprehensive study of the properties of engineering materials.)

Johnson, C. G.: "Metallurgy," American Technical Society, Chicago, 1952. (A great deal about the manufacture and behavior of metals.)

Perry, J.: "The Light Metals in Industry," Longmans, Green & Co., Inc., New York, 1947. (A good survey for the uses of these metals.)

Shrager, A. M.: "Elementary Metallurgy and Metallography," The Macmillan Company, New York, 1949. (Basic principles simply explained. A unique combination of principles and practices.)

APPLIED MECHANICS AND MECHANISMS

Breneman, J. W.: "Mechanics," McGraw-Hill Book Company, Inc., New York, 1948. (An elementary text. Useful applications of the simple machines.)

Church, A. H.: "Kinematics of Machines," John Wiley & Sons, Inc., New York, 1950. (Fundamental analyses of the motions of mechanisms.)

Cox, G. N.: "Engineering Mechanics," D. Van Nostrand Company, Inc., New York, 1954. (Comprehensive treatment on an advanced level.)

Jensen, A.: "Applied Engineering Mechanics," McGraw-Hill Book Company, Inc., New York, 1947. (An elementary text. Many good problems. Well illustrated.)

Winston, S. E.: "Mechanisms," American Technical Society, Chicago, 1941. (How machine parts transmit and modify motion.)

DESIGN OF MACHINE PARTS

Berard, S. J., and E. O. Waters: "The Elements of Machine Design," D. Van Nostrand Company, Inc., New York, 1954. (Good descriptions and illustrations of various machine parts.)

Black, P. H.: "Machine Design," McGraw-Hill Book Company, Inc., New York, 1948. (A comprehensive text. Advanced.)

Bradford, L. S., and P. B. Eaton: "Machine Design," John Wiley & Sons, Inc., New York, 1947. (Good illustrations. A very full treatment of the topic of bearings. More elementary level.)

Faires, V. M.: "Design of Machine Elements," The Macmillan Company, New York, 1941. (A comprehensive and very readable text with good design analyses. Excellent illustrations. Advanced.)

Hyland, P. H., and J. B. Kommers: "Machine Design," 3d ed., McGraw-Hill Book Company, Inc., New York, 1943. (Combines shop processes, materials of industry, some kinematics, and applied mechanics with the design of machine parts. Advanced.)

Lipson, C., G. C. Noll, and L. S. Clock: "Stress and Strength of Manufactured Parts," McGraw-Hill Book Company, Inc., New York, 1950. (Analyses of various conditions of fatigue and stress concentration in manufactured parts.)

Maleev, V. L.: "Machine Design," International Textbook Company, Scranton, Pa., 1946. (An excellent text on an advanced level. Very comprehensive in scope with thorough design analyses. Good illustrations.)

Marks, L. S.: "Mechanical Engineers' Handbook," 5th ed., McGraw-Hill Book Company, Inc., New York, 1951. (A recognized standard handbook for all branches of mechanical engineering.)

Nachman, H. L.: "Elements of Machine Design," John Wiley & Sons, Inc., New York, 1918. (An elementary text. Good for fundamentals.)

Salisbury, J. K. (editor): "Kent's Mechanical Engineers' Handbook," John Wiley & Sons, Inc., New York, 1950. (A recognized standard handbook for all branches of mechanical engineering.)

Spotts, M. F.: "Design of Machine Elements," Prentice-Hall, Inc., New York, 1953. (Includes thorough stress analyses for machine parts. Advanced.)

Vallance, A., and V. L. Doughtie: "Design of Machine Members," 3d ed., McGraw-Hill Book Company, Inc., New York, 1951. (Good illustrations and good coverage on an advanced level.)

Winston, S. E.: "Machine Design," American Technical Society, Chicago, 1940. (Emphasis on thorough and simple derivations of machine-design formulas. More elementary.)

VISUAL AIDS

The visual aids listed below and on the following pages can be used to supplement much of the material in this book. For the convenience of users the films have been grouped under three headings: (1) properties of metals and industrial processes, (2) applied mechanics and mechanisms, and (3) design of machine parts. We recommend that each film be reviewed before use in order to determine its suitability for a particular group or unit of study.

Motion pictures and filmstrips are included in the following list, the character of each being indicated by the abbreviations "MP" and "FS." Immediately following this identification is the name of the producer and, if different from the producer, the name of the distributor. Abbreviations are used for these names and are identified in the list of sources at the end of the list. Unless otherwise indicated, the motion pictures are 16-mm sound black-and-white films and the filmstrips are 35-mm black-and-white silent films. The length of motion pictures is given in minutes (min), of filmstrips in frames (fr).

Most of the films can be borrowed or rented from state and local film libraries. A list of these sources is given in *A Directory of 2660 16-mm Film Libraries*, available for 50 cents from the Government Printing Office, Washington 25, D.C.

This bibliography is a selective one, and film users should also examine the latest annual edition and supplements of *Educational Film Guide* and *Filmstrip Guide*, published by the H. W. Wilson Co., New York. The *Guides*, standard reference books, are available in most school, college, and public libraries.

Properties of Metals and Industrial Processes

Add Power to Your Hands (MP, Mohawk, 10 min color or BW)   Describes the processes of drop forging, heat-treating, precision hardening, and electronic induction hardening as applied to the manufacture of pliers.

Hardness Testing: Rockwell (MP, USOE/UWF, 18 min)   Need for hardness testing; how to set up the Rockwell hardness tester; select and seat the penetrator, select and mount the anvil, adjust the timing of the machine, and test flat and curved surfaces.   (Correlated filmstrip, 49 fr)

Heat Treatment of Aluminum, Parts 1 and 2 (MP, USOE/UWF)

Part 1 (19 min)   Purpose of heat-treatment, microstructure changes during heat-treatment, procedure of heat-treatment, aging or precipitation hardening, and effects of heat-treatment on the physical properties of aluminum.   (Correlated filmstrip, 47 fr)

Part 2 (24 min)   Nature of cold-working operations; microstructure changes during cold working and during annealing; cold-working and annealing operations.   (Correlated filmstrip, 41 fr)

Heat Treatment of Steel (MP-FS series, USOE/UWF)   Three motion pictures with correlated filmstrips as follows:

Elements of Surface Hardening (14 min)   Shows how steel is packed and gas-carburized; how a thin, hard case is obtained by cyaniding; how nitriding is used to obtain a very hard case; and how steel is flame- and induction-hardened. (Correlated filmstrip, 36 fr)

Elements of Hardening (15 min)   How steel is quench-hardened; how the structure and hardness of steels with different carbon content change at progressive quench-hardening stages; how the lower and upper critical temperatures of steel are determined; and how an iron-carbon diagram is constructed and what it shows.   (Correlated filmstrip, 40 fr)

Elements of Tempering, Normalizing, and Annealing (15 min)   How steel is tempered; how the structure, toughness, and hardness of plain carbon steel change at progressive tempering stages; how steel is normalized and annealed; and the results of normalizing and annealing.   (Correlated filmstrip, 31 fr)

Induction Heat—Industry's Modern Tool (MP, AC, 20 min color)   Shows production techniques used in the installation of induction heaters including brazing, annealing, hardening, and soft soldering.   Intended for engineering groups interested in the heat-treating of metals.

Iron-Carbon Alloys (MP, ASM, 30 min color)   Points out the varying properties of these alloys and their uses in industry; describes the tools of metallurgical research; and shows by animation and photomicrographs the formation and transformation of structures in iron-carbon alloys as they are heated and cooled.

Metal Crystals (MP, ASM, 33 min silent)   Defines and illustrates crystalline and noncrystalline substances; illustrates microscopic techniques with metal specimens; and explains the metal structures in terms of their cooling characteristics. (Sponsored by the American Society for Metals and the Ohio State University Research Foundation.)

Metal Magic (MP, GE, 10 min)   Shows the crystalline structure of metals and the development of new (1945) alloys.   (Excursions in Science series.)

Potomac Hot Die Steels (MP, AL, 20 min color)   Metallurgical explanation of the various properties and applications of hot-work tool steel.

Powder Metallurgy, Parts 1 and 2 (MP, USOE/UWF)

Part 1, Principles and Uses (19 min)   Principles of powder metallurgy—powder, pressure, heat; major industrial applications of powder metallurgy; laboratory process of combining silver and nickel powders.

Part 2, Manufacture of Porous Bronze Bearings (15 min)   Manufacturing process by which metal powders are fabricated into porous bronze bearings and impregnated with oil.

Science of Making Brass (MP, Chase, 29 min color)   Shows the process of making brass and copper alloys and forming sheets, rods, wire, and tubes.

Tension Testing (MP, USOE/UWF, 21 min)   How a hydraulic tension-testing machine operates; how to prepare the machine and a specimen for a test and conduct the test to determine the specimen's elastic limit, yield point, and ultimate strength.   (Correlated filmstrip, 45 fr)

This Moving World (MP, Assn, 30 min color)   Shows the preparation of molds, melting of metal, and pouring of molten iron into the molds of an iron foundry; describes the annealing process; and illustrates various applications of malleable iron.   (Sponsored by the Malleable Founders Society.)

Applied Mechanics and Mechanisms

An Introduction to Vectors: Coplanar Concurrent Forces (MP, USOE/UWF, 22 min) Explains the meaning of scalar and vector quantities; how to add scalars and vectors; methods of vector composition and vector resolution; relationship between vector composition and vector resolution; and how vectors may be used to solve engineering problems. (Correlated filmstrip, 36 fr)

A Pictorial Study of Methods Improvement Principles (MP, Saginaw, 45 min color silent) Shows such principles as operation combination, automatic ejection, multiple processing, and barrier elimination on typical grinding, drill-press, punch-press, assembly, heat-treating, and packaging operations.

Principle of Moments (MP, USOE/UWF, 23 min) Explains the concept of moment of a force, the formula for finding its numerical value, and principle of moments as applied to all coplaner force systems. (Correlated filmstrip, 29 fr)

Principles of Dry Friction (MP, USOE/UWF, 17 min) Defines friction; explains the advantages and disadvantages of friction, the forces involved in friction, static and kinetic friction, and the calculation of the coefficients of static and kinetic friction. (Correlated filmstrip, 36 fr)

Principles of Gearing: An introduction (MP, USOE/UWF, 18 min) Friction gears and toothed gears; law of gearing, positive driving, involute profiles, pressure angle, cycloid profiles, velocity rates, and circular pitch. (Correlated filmstrip, 37 fr)

Principles of Lubrication (MP, USOE/UWF, 16 min) The need for lubrication, properties of lubricants, action of lubricants, viscosity of lubricants, and conditions that determine proper viscosity. (Correlated filmstrip, 33 fr)

Transmission of Rotary Motion (MP, YAF, 11 min) Explains how power is transmitted from one point to another by means of shafts, gears, belts, and chains.

Design of Machine Parts

Advance Guard (FS, Sheffield, 50 fr with disk recording, $33\frac{1}{3}$ rpm, 15 min) Describes the many and varied services performed by the manufacturers of precision gages for industry. For college, industrial, and engineering groups.

Art of Generating and Gear Manufacturing Equipment (MP, FG, 15 min color) Explains the theory, design, tooth action, and contact of gears and their production and inspection.

Building Motion Economy into Tools and Equipment—Some Examples and Suggestions (MP, Saginaw, 60 min silent) Shows how machines and equipment can be better designed to enable an operator to produce more pieces with less effort.

Engineering Drawing (MP-FS series, McGraw) Ten motion pictures and follow-up filmstrips correlated with French and Vierck, *Engineering Drawing*. Titles are:

According to Plan: Introduction to Engineering Drawing (9 min)
Orthographic Projection (18 min)
Auxiliary Views: Single Auxiliaries (23 min)
Auxiliary Views: Double Auxiliaries (13 min)
Sections and Conventions (15 min)
The Drawings and the Shop (15 min)

Selection of Dimensions (18 min)
Pictorial Sketching (11 min)
Simple Developments (11 min)
Oblique Cones and Transition Developments (11 min)

Fitting and Scraping Small Bearings (MP, USOE/UWF, 20 min)   How to scrape split bearings for shaft fit and alignment, "relieve" a split bearing to aid lubrication, and cut oil grooves in the cap of a split bearing.   (Correlated filmstrip, 50 fr)

Holding Power (MP, Bethlehem, 25 min color.)   Shows the manufacture and uses of bolts, nuts, and allied steel fasteners.   Includes scenes of automatic bolt-making machines and nut formers in operation, laboratory testing, packaging, and warehousing operations.

Micro Instrument Ball Bearings, Their Care and Handling (MP, GM, 22 min) Technical film for engineers showing the care taken by General Motors in producing ball bearings and the equally conscientious care which users should give.

Mighty Miniatures (MP, Miniature, 15 min color)   Describes the manufacture and inspection of miniature bearings and their applications in precision mechanisms.   Scenes show how bearing rings are machined and polished and how the completed assembly is tested for concentricity, torque, etc.

Oil Films in Action (MP, GM, 18 min color)   Technical film prepared for practicing engineers, designers, and students to show how oil films behave in bearings.

Film Sources

AC—Allis-Chalmers Mfg. Co., Milwaukee 1, Wis.
AL—Allegheny Ludlum Steel Corp., 2020 Oliver Bldg., Pittsburgh 22.
ASM—American Society for Metals, 7301 Euclid Ave., Cleveland 3.
Assn—Association Films, Inc., 347 Madison Ave., New York 17.
Bethlehem—Bethlehem Steel Co., Bethlehem, Pa.
Chase—Chase Brass and Copper Co., Waterbury 20, Conn.
FG—Fellows Gear Shaper Co., 78 River St., Springfield, Vt.
GE—General Electric Co., Schenectady, N.Y.
GM—General Motors Corp., 3044 W. Grand Blvd., Detroit 2, or 405 Montgomery
    St., San Francisco 4.
McGraw—McGraw-Hill Book Company, Inc., 330 West 42d St., New York 36.
Miniature—Miniature Precision Bearings, Keene, N.H.
Mohawk—Mohawk Productions, 125 Mayro Bldg., Utica, N.Y.
Saginaw—Saginaw Steering Gear Division, General Motors Corp, 628 N. Hamilton
    St., Saginaw, Mich.
Sheffield—Sheffield Corporation, Dayton 1, Ohio.
USOE—U.S. Office of Education, Washington 25, D.C.
UWF—United World Films, Inc., 1445 Park Ave., New York 29.
YAF—Young America Films, Inc., 18 E. 41st St., New York 17.

INDEX

Allowable bearing pressure, 137
Allowance, 208
Alloy bronze, 24, 25
Alloys, 17
Aluminum, 25
  alloys, 25
  bronze, 24
American Iron and Steel Institute, (AISI), code for steels, 22
American Society of Mechanical Engineers (ASME), classes of fits, 208
  combined stresses in shafts, 80
  effect of keyways on design stress, 93
  variable loading in shafts, 56, 57
Annealing, 18

Babbitt metal, 26
Ball bearings (*see* Bearings, rolling-contact)
Beams, equilibrium in, 75
  maximum bending moment, 77–79
  reactions, calculation of, 75–79
  shear diagrams, 76–78
  vertical shear in, 76–79
  zero shear in, 77–79
Bearing, stress of, in bearings, 136, 137
  in couplings, 130
  in keys, 94, 95
  in riveted joints, 160
Bearings, classifications, 135, 136
  rolling-contact, 136, 140–142
    fatigue in rollers, 142
    lubrication, 142
    materials, 142
    parts, 141
  sliding-contact, 135–140
    allowable bearing pressure, 137
    bushings, 138
    design, 136–138
    friction, 138–139
    heat generated, 139
    lubrication, 135, 137, 138
    materials, 138, 139
  step, 143
  thrust, 142, 143
Belt tensions, 74–75, 108

Belting, 106–110
  angle of contact, 107
  design, 107–110
  design stress, 107
  efficiency, 107
  friction, 108, 109
  materials, 106
  slippage, 108
  splices, 107
  tensions, 108
Bending, stress of, arm pulley, 103–106
  columns, 201–205
  curved beams, 191–193
  gear teeth, 119–121
  shafts, 80–84
  springs, 182–185, 187, 188
Boilers and pipes, joint design, riveted, 162–167
  welded, 157
  materials, 163
  pressure analysis, 154, 155
  wall design, 155, 156
Bolts (*see* Screws)
Brass, 22, 23
  cartridge, 22, 23
  Muntz metal, 23, 24
  red, 22, 24
Brittleness, 12
Bronze, alloy, 24, 25
  aluminum, 24
  gun metal, 24
  nickel, 24
  phosphor, 24
Bushings, 25, 138, 139

C clamp, 193
Cap screw, 169, 170
Carbon, in casehardening, 18
  in cast iron, 18
  in steel, 20, 21
Carburizing, 18
Casehardening, 18, 21
Cast iron, gray, 18, 19
  malleable, 18, 19
  white, 18, 19
Caulking joints, 164

229

Center-of-gravity axis, curved beam
    sections, 192n.
  unsymmetrical sections, 194, 195
Centrifugal force, in belts, 107
  in flywheels, 52, 53
Chilled castings, 19
Clutches, friction, cone, 146–151
    clutch angle, 151
    design, 147–151
    frictional resistance, 147, 148
    materials, 150
    power transmission, 149
    pressure on contact area, 150
   disk, 145, 146
  function, 145
  positive, spiral-jaw, 152
    square-jaw, 152
Coil springs, 181, 186–188
Cold-rolled steel, 21
Cold working, 17
Collar bearing, 143
Columns, average stress, 203
  circular section design, 205
  combined with beam loadings, 204
  critical load, 201
  eccentrically loaded, 204
  Johnson formula, 203
  radius of gyration, 203
  slenderness ratio, 203
  stress analysis, 201, 202
Combined stresses, compression and
    bending, 201–205
    stress analysis, 201, 202
  tension and bending, 191–201
    stress analysis, 191–193, 196, 197
Copper, 22
Coupling bolt, 169
Couplings, flange, 128
  muff, 128
  Oldham, 132, 133
  safety flange, 128, 129
    design, 129–132
  universal joint, 133
Crane, force analysis, 32, 33
Crane hook, 191–194, 199–201
  stress analysis, 192
Creep, 15
Creep strength, 15
Critical load (in columns), 201
Critical range, 17
Critical temperatures (points), 17

Dangerous section, in beams, 77
  in columns, 202
  in curved beams, 191

Deflection, in shafts, 87–91
  in springs, 188–190
Design stress, 50
Disks in step bearings, 143
Dow metal, 25
Drawing back, 18
Ductility, 12
Duralumin, 25
Dynamic forces, 52

Efficiency, joints, belt, 107
    riveted, 156, 160–162
    welded, 156, 158
  simple machines, 43–45
Elastic limit (proportional limit), 9
Elasticity, 11
Endurance limit, 14
Equilibrium, 27
  conditions for, 29
  and motion, 33–35
Equivalent bending moment, 81
Equivalent twisting moment (torque), 80
Extensometer, 7

Factor of safety, 50, 51
  equation, 51
Fatigue, endurance limit, 13, 14
Fatigue failure, 13, 14
Fatigue strength, 13, 14
Fit in screws, 173
Fits, force and shrink, frictional resist-
    ance, 210, 211
  stress analysis, 209, 210
Flange coupling (safety flange coupling),
    128–132
Flat springs, 181–183
Flexure formula, general, 81
  rectangular beams, 119, 120
Force fit (see Fits)
Forces (loads), centrifugal, 52, 53, 107
  components, 27, 28
  composition, 27
  concurrent, 27
    analytical solutions, 29–38
    graphical solutions, 28, 29
  dynamic, 52
  elastic, 11, 12
  equilibrant, 27, 28
  gear, 73–75
  impact, 53, 56, 57
  inertia, 52
  polygon, 29
  pulley, 73–75
  repeated applications of, 13, **14**
  repeated reversals, 13, 14
  resolution, 27

Forces (loads), resultant, 27, 28
Friction, in bearings, 138, 139, 141
  in belts, 109
  coefficient of, rolling, 141
    sliding, 139
  in friction clutches, 147, 148
  oil, internal, 139
  sliding, analysis, 108, 109
Friction wheels, 124, 125
Fusion welding, 157

Gage line (in riveted joints), 163
Gear teeth, circular pitch, 115
  design stresses, 122
  diametral pitch, 115
  Lewis equation, 123, 124
  parts of, 114, 115
  principal design proportions, 121
  relationships of parts, 115, 116
  sizes and diametral pitch, 116
  spur, design of, 119–121
  wear, 124
Gear train, efficiency, 43–45
  practical mechanical advantage, 45
  theoretical mechanical advantage,
    43–45
Gears, bevel, 113
  helical (spiral), 112
  involute system, 118, 119
    base circle, 118
  materials, 113
  miter, 113
  pinion, 113
  pitch circle, 114
  pressure angle, 123
  spur, 112
  tooth thrust, 73–75
  worm, 113
Gland, 179
Graphite (in gray cast iron), 18
Gun metal, 24

Handbooks, engineering, 5, 6
Hardenability, 21
Hardness, 13
Heat-treatment, 17
Horsepower, 58, 59
  equation, 58
Hot-rolled steel, 21
Hydraulic jack, lifting capacity, 40, 41
  Pascal's principle, 41
  pressure in, 41

Ideal torque, 81
Impact tests, Charpy, 15
  Izod, 15

Impact tests, tension, 15
Inertia loads, 52, 53
Interference, 208
Iron carbide, in casehardening, 18
  in cast iron, 19
  in steel, 20

Joints, gasket, 177, 178
  ground, 177, 178
  riveted (see Riveted joints)
  shrink and force (see Fits)
  welded (see Welded joints)
Journal, 135, 136

Keys, feather, 99
  flat and square, design, 94–96
    sizes, 96
  materials, 94
  pin, 98
  saddle, 98
  splines, 99
  Woodruff, sizes, 98
    uses, 97, 98

Leaf springs, 184, 185
Lewis equation for gear teeth, 123, 124
Loads (see Forces)
Locking devices, 170

Machinability, 13
Machine bolt, 169
Machine design, definition, 1, 3
  design procedure 3, 56, 57
Machine screw, 170
Machines, compound, gear train, 43–45
  screw jack, 41–43
  simple, hydraulic jack, 40, 41
    lever, 1, 2
    pulley, 1, 2
    wheel and axle, 1, 2
Malleability, 13
Malleable iron, 18, 19
Manufacturers' catalogues, 3, 4
  for clutches, 152
  for couplings, 131
  for pulleys, 4
Mechanical advantage, practical, 44
  theoretical, 41, 42
Mechanisms, definition, 1
  examples, 2, 5
Modulus of elasticity, 10, 11, 86
Modulus of resilience, 11, 12
  from stress-strain diagram, 12
Modulus of rigidity, 86

Moment of inertia, symmetrical sections, about an axis (rectangular), 81, 215
  about a point (polar), 67
  unsymmetrical sections about an axis (rectangular), 195, 196, 215
Monel metal, 25
Muff coupling, 128

Newton's second law, applications of, 34, 45, 46, 53
Nickel bronze, 24
Normalizing, 18

Oldham coupling, 132, 133

Pascal's principle, 41
Phosphor bronze, 24, 25
Pitch, gears, circular, 115
  diametral, 115
  riveted joints, 163
  screw threads, 172
Pitch circle, 114
Power, 58–63
  relationship, to speed and force, 59
  to speed and torque, 59–61
  transmission through shafts, 62, 63
Pressure, hydraulic press, 41
  pressure vessels (boilers and pipes), 154, 155
  shrink and force fits, 209, 210
Pressure angle (gears), 123
Pressure vessels, cylindrical, 154–157
  spherical, 156
Progressive fracturing, 13, 14
Properties of materials, 7
  temperature effect, 15
Proportional limit (elastic limit), 9
Pulleys, arm, 103–106
  belt pull, 73–75
  materials, 101
  selection, 106
  solid-web, 102, 103
  types, 101, 102
Punch press, frame investigation, 193, 198, 199
  kinetic energy, 45–47
  shearing force, 47

Radius of gyration, 203
Reactions, in beams, 75
  in shafts, 136
Resilience, 11, 12
  modulus of, 11, 12
Riveted joints, boiler joint design, 164–167

Riveted joints, caulking, 163, 164
  design stresses, 161
  efficiency, 160–162
  failure possibilities, 159
  parts, 163
  safe loads, 160–162
  types, 158
Roller bearings (see Bearings, rolling-contact)

Safety flange coupling, 128–132
Screw bracket, 176, 177
Screw jack, lifting capacity, 41, 42
  theoretical mechanical advantage, 43
Screw threads, acme, 173
  American National Standard, listing, 172
  parts, 171, 172
  buttress, 173
  SAE Extra Fine, 173
  square, 173
Screws, 169–176
  design, 174–176
  design stresses, 175
  materials, 170
  for power transmission, 173, 174
  stresses in, 174
  types, 169, 170
Seals, gasket joints, 177, 178
  ground joints, 177, 178
  stuffing boxes, 178, 179
Shafts, as beams, 73–79
  elastic deformation, 86–91
  fatigue, 79
  hollow, design (bending and torsion), 80–84
    design (torsion only), 69, 70
    sizes, 70, 71
  reactions, 73
  solid, design (bending and torsion), 80–84
    design (torsion only), 68
    materials, 66, 67
    sizes, 67
  torsional deflection, 87–89
    equations, 88
    limits, 89
  transverse deflection, 90–91
Shear, stress of, in couplings, 130
  in keys, 94, 95
  in riveted joints, 160
  in screws, 174, 175
  in solid-web pulley, 102, 103
Shear modulus of elasticity, 86
Shrink fit (see Fits)
Silicon in cast iron, 18

Slag, 20
Slenderness ratio, 203
Society of Automotive Engineers (SAE),
    clutch angle recommendations, 151
    code for steels, 22
    Extra Fine screw threads, 173
Softening, 17
Solid solution, 18
Speed relationship, to power and force,
    59
    to power and torque, 59–61
Splines, 99
Spring scale, 188
Springs, coil, design, tension and com-
    pression, 186, 187
        torsion, 187
    deflection, 188–190
    flat, design, 182, 183
    leaf, construction, 185, 186
        design, 184, 185
    materials, 25, 181
    scale, 188
    types, 181
Sprockets, 2, 112n.
Steam engine, reciprocating, force
        analysis, 35
    force on wrist pin, 54
Steel, alloy, 21, 22
    plain carbon, 20, 21
Step bearing, 143
Stiffness, 11
Stove bolt, 170
Strain, definition, 7
    total, 7
    unit, 7
Strength, 12
Stress, allowable, 50
    definition, 7
    design, 50
    induced, 50
    limit, 50
    repeated, 13, 14
    residual, 18
    reversals, 13, 14
    total, 7
    unit, 7
Stress concentration, 54–56
    factor, 56
    in keyways, 93

Stress concentration, in screws, 175
Stress-strain diagram, 8-12
Stud bolt, 169, 176
Stuffing boxes, 178, 179
    gland for, 179

Tempering, 18
Tension, stress of, in pressure vessels, 155
    in riveted joints, 160
    in screws, 174
    in welded joints, 158
Tension test, 7-11
    necking down in, 9, 10
Through bolt, 169
Thrust bearings, 143
Tin in bronze, 24
Toggle joint, force analysis, 36–38
Tolerances, 211, 212
    definition, 212
    mating parts, 212
    nonmating parts, 212
Torque (twisting moment), relationship
        to power and speed, 59
    varying values in shaft, 63, 64
Torque wrench, 174n.
Torsion, formula, 67
    resisting moment of, 67
    stress of, in screws, 174
        in shafts, 66
Toughness, 12
Twisting moment (see Torque)

Ultimate strength, 9, 51
Universal joint, 133

Washers, 171
Welded joints, design stresses, 158
    efficiency, 158
    fusion welding, 157, 158
    types, 158
Work, definition, 42
    relationship to power, 58
Wrought iron, 19, 20
    slag, 20

Yield point, 9

Zinc in brass, 23